DIGGING UP THE ICE AGE

Recognising, recording and understanding fossil and archaeological remains found in British quarries

A GUIDE AND PRACTICAL HANDBOOK

Simon Buteux, Jenni Chambers and Barbara Silva

Edited by
Simon Buteux

with contributions by

David Keen, Danielle Schreve and Mark Stephens

Published by

Archaeopress
Gordon House
276 Banbury Road
Oxford OX2 7ED
England
www.archaeopress.com

DIGGING UP THE ICE AGE
Recognising, recording and understanding fossil and archaeological remains found in British quarries

A GUIDE AND PRACTICAL HANDBOOK

© the authors and the University of Birmingham 2009

ISBN 978-1-905739-24-0

Printed in England by Information Press

All rights reserved. No part of this publication may be reproduced, stored in retrieval system, or transmitted, in any form or by any means, electronic, mechanical, photocopying or otherwise, without the prior written permission of the copyroght owner

Contents

Acknowledgments	v
Chapter 1: Introduction – Quarrying and The Ice Age	1
Chapter 2: Quaternary Science and Climate Change	5
Chapter 3: Human Evolution and Ice Age Britain	23
Chapter 4: What Remains? – Sediments	52
Chapter 5: What Remains? – Fossils	65
Chapter 6: What Remains? – Archaeology	88
Chapter 7: How Old Is It? – Dating the Ice Age	104
Chapter 8: A Guide to Recording Ice Age Sites	111
Glossary: An Ice Age Dictionary	121
Further Reading	164
Useful Organisations and Websites	172
Appendix 1: Example of Sediment Description Sheet	176
Appendix 2: Example of Sample Record Sheet	177
Illustration Sources and Credits	178
Index	185

The preparation and publication of this book has been supported by a grant from the Aggregates Levy Sustainability Fund through grant schemes administered by English Heritage and Natural England.

Acknowledgments

The writing of this handbook was part of a project called the National Ice Age Network (NIAN). NIAN was funded (2004-7) by grants from the Aggregates Levy Sustainability Fund (ALSF) jointly administered by English Heritage and Natural England. The purpose of the ALSF is to help address and mitigate the environmental costs of aggregate extraction, and to promote sustainable approaches. The main purpose of NIAN was to raise awareness of the importance and interest of the Ice Age remains that are often uncovered during aggregate extraction, and to promote approaches that would ensure that this material was recognised, recorded, recovered and reported. This handbook is a contribution to that purpose.

The authors would like to thank Buzz Busby, Kath Buxton and Ingrid Ward at English Heritage, and Natalie Bennett and David Hodson at Natural England, for their support and advice during the NIAN project. We are also very grateful to Jonathan Last and Ingrid Ward (English Heritage), Rob Hosfield (University of Reading), Clive Gamble and Danielle Schreve (Royal Holloway, University of London) for reading and commenting on earlier drafts of the book; of course they are not responsible for the faults that remain or the opinions expressed. We would also like to thank Rebecca Beardmore for steering the manuscript through to publication and for securing permission where required for the use of illustrations.

Particular thanks are due to all the quarry companies, quarry managers and quarry workers across the country who have assisted in many ways to ensure that Ice Age remains uncovered during quarrying have been recovered, and who for generations have collaborated with researchers studying these remains. Although the example of good practice set by these companies and individuals has unfortunately not always been followed by others, they show how much can be achieved and how rewarding it can be for all concerned.

During the course of the NIAN project its director, Professor David Keen, died of cancer. David devoted much of his distinguished career to the study of past environments through remains uncovered by quarrying (his particular speciality was snails), and by personal example did much to promote mutual understanding and good relations between researchers and the industry. He kept working – and advocating the importance and interest of the Ice Age – almost to the end of his illness. This book is dedicated to him.

Chapter 1: Introduction – Quarrying and The Ice Age

For over a hundred years, sand and gravel quarrying has been of enormous benefit to geology, palaeontology and archaeology – quarries have been the main source of Ice Age fossils and finds. It is because of deep excavations into Ice Age sediments that the geological sequences, the fossil remains of plants and animals, and the stone tools of Britain's earliest human inhabitants have come to light (Fig. 1).

Fig. 1: A quarry man and a specialist (Professor David Keen, foreground) inspect fossil remains uncovered by quarrying.

These discoveries are of more than just narrow academic interest and importance. They provide much of the information needed to reconstruct the changing patterns of plants, animals and human occupation in Britain, and the development of the landscape itself. Furthermore, the Ice Age was a period that saw great changes in climate and environment. It is only by understanding such changes in the past that we can properly understand them in the present.

In this important endeavour there has long been a partnership, largely informal, between researchers and many in the quarry industry. This handbook describes something of what has been achieved. However, quarrying is also a destructive process. There is a danger, if the issues are not understood and the appropriate steps are not taken, that much important evidence may be lost without record. The benefits of aggregate quarrying to understanding and reconstructing the Ice Age world cannot be realised unless the fossils and archaeological remains are recorded and recovered.

This handbook is aimed at all non specialists involved with or interested in quarrying and the contribution it makes to understanding our deep past. This includes archaeologists and geologists whose specialism lies in other periods as well as those working in the industry itself. It assumes no prior knowledge and technical terms are introduced as the need for them arises, with an extensive glossary provided at the end of the book. The glossary – an 'Ice Age dictionary' – includes not only explanations of the terms that are used in the book but also definitions and brief explanations of terms and techniques that will be encountered more widely in the literature about the Ice Age.

The main aim of this handbook is practical. Following the introductory chapters, in chapters 4 to 7 we describe the types of evidence that can be found, how that evidence is recovered and studied, and what it has the potential to tell us. In chapter 8 we provide guidelines on the simple procedures that can be followed in a typical quarrying situation to ensure that the maximum amount of this evidence can be recognised, reported, recorded and recovered. As will be seen, these procedures are very different to those typically associated with an archaeological excavation, with which they should not be confused. They are generally more of the character of a geological investigation, are typically low cost, and seldom involve significant disruption to quarrying.

However, before moving on to these practical issues the remainder of this chapter and chapters 2 and 3 are concerned with providing some of the necessary background.

Why is the Ice Age Important?

The landscape of the British Isles has been shaped by Ice Age events and forces. For more than two million years the Earth's climate has cycled repeatedly between cold, glacial periods and warmer, interglacial periods. These cycles of climate change have left their mark on the landscape around us.

Around 450,000 years ago Britain looked very different from today. At this time an ice sheet about a mile thick covered much of the North and Midlands, reaching what is today the outskirts of London. Beyond the ice sheet the landscape would have been bleak and treeless, with very few animals and plants able to survive the extreme temperatures and harsh conditions. Even when the ice sheets were not at their maximum, conditions were, for tens of thousands of years, much colder than they are today (Fig. 2).

Move forwards 50,000 years to around 400,000 years ago, and the climate had changed dramatically. The climate conditions were fairly similar to today although the animals were different (Fig. 3). In the Thames Valley you could have seen straight-tusked elephants and forest rhinoceros being hunted by lions. You might even have seen some of our early ancestors making stone tools as they moved along the newly-formed valley.

What we have described is just half of one long climate cycle from cold to warm and back to cold again.

Fig. 2: The British landscape during a glacial period. The floodplain of a braided river in periglacial conditions. Mammoths and wolves are in the foreground, with a red deer by the river and a woolly rhinoceros on the opposite bank.

Fig. 3: The British landscape during an interglacial period. A bear advances towards the river, while two spotted hyaenas feed on a wild boar, with a giant deer in the water behind them. Giant and fallow deer are on the other side of the river, while in the background are straight-tusked elephants.

Such cyclic climate change is the key characteristic of the Ice Age and resulted in ice sheets expanding and retreating over much of the northern hemisphere. It is against this backdrop of climate change that humans evolved.

For over two centuries scientists have been striving to understand the complex mechanisms of Earth history and climate change. This information is vital if we are to understand how British environments have evolved and changed through time, and how they will change again in the future.

A Note on Dates

All the dates mentioned in this book are given in years Before Present, often abbreviated as 'BP'. The dates are based on various scientific dating methods, some of which are briefly described in chapter 7. The most important thing to bear in mind about these dates is that they are approximations, and 'leeway' of thousands or sometimes tens of thousands of years should be allowed.

As a general rule, the older the date the more approximate it is. In many respects this imprecision doesn't matter that much – if one is talking about a time 450,000 years ago, a few thousand years this way or that is often of no real significance. However, as one approaches the present day, the degree of imprecision begins to matter more.

Many dates cited in this book that fall in the period between about 50,000 BP and 11,400 BP (the latter being the conventional date for the end of the Ice Age) are based ultimately on the radiocarbon dating method. As explained in more detail in chapter 7, 'raw' radiocarbon dates need to be 'calibrated' to give an approximation of the real, calendar date. Unfortunately for the period in question towards the end of the Ice Age (50,000–11,400 BP), there is no totally reliable and generally agreed way of calibrating radiocarbon dates. Many archaeologists therefore prefer to use uncalibrated dates, which may be as much as 2,000–3,000 years 'too young' compared to real, calendar dates. Although this practice is internally consistent, it means that these archaeologists are 'out of sync' with most other scientists.

Climatologists, for example, do not need to rely on radiocarbon dating and can get some very accurate dates for climate events from ice-cores (see chapter 2).

In order to maintain consistency throughout the book and attempt to avoid confusion, all the dates we cite are calendar dates BP, and if based on radiocarbon determinations they are therefore calibrated. Because of this, readers may note a discrepancy of up to about 3,000 years in the dates we cite and dates to be found in some of the recent archaeological literature. Nevertheless, the uncertainties about the reliability of the radiocarbon calibration methods available for this period must be acknowledged. This makes it important to stress that the dates we give should be treated as indicative only. Much progress is being made in the field of scientific dating and it is hoped that in the future it will be possible to cite dates with much more confidence.

Finally, there is of course a difference of about 2,000 years between dates BP, as used throughout this book, and dates BC ('Before Christ'). Most of the Ice Age is so remote in time that this makes no real difference, but readers should be aware of this difference where dates towards the present day are cited.

Chapter 2: Quaternary Science and Climate Change

What is the Quaternary?

Like all scientific subjects, the study of the Ice Age involves its own specific terminology. This terminology has developed over many years, during which ideas about the history of the Earth, and particularly our understanding of its duration, have changed radically. Many terms still in use betray their origins in now outdated concepts. The result can be confusing, even for the specialist. Here we attempt to introduce only those terms as will be commonly encountered.

One technical term for the Ice Age is the 'Quaternary Period' and the study of this period, involving many different disciplines, is called Quaternary science. 'Quaternary' literally means 'fourth' and it refers to the most recent period of Earth history – we are still in it – from a geological perspective. In geological terms it is a very short period and the implication that there were only three significant periods before it is now very antiquated, stemming from a time when the Earth was thought to be a few thousand, rather than a few billion, years old. Nevertheless, misleading though it perhaps is, the term has been retained (Fig. 4).

Dating the onset of the Ice Age or Quaternary Period is difficult and subject to some controversy. However, a date of about 2.6 million years ago is widely accepted. The Quaternary Period is sub-divided into the Pleistocene ('most recent') and the Holocene ('wholly recent') epochs. The Pleistocene is the period up to around 11,400 years Before Present (BP). The Holocene refers to our present interglacial warm period, from around 11,400 BP to today. For many purposes the terms 'Quaternary' and 'Pleistocene' can be used interchangeably, and both terms will be used throughout this book along with the more informal 'Ice Age'. Some writers use the term 'ice age' to refer to the glacial periods within the Ice Age or Quaternary as a whole. We avoid this usage.

There is little reason not to believe that the current interglacial will come to an end and that the Earth will enter a glacial period again in the future, if humanly-caused global warming does not upset the natural rhythms. In this sense, we are still living in the Ice Age.

Glacial Cycles and Their Causes

The Pleistocene is characterised both by major climate cycles, from glacial to interglacial and back to glacial again, and by dramatic climate fluctuations over shorter timescales. These climatic changes have varied in severity and duration. From a little after one million years ago whole major cycles have tended to last about 100,000 years. The glacial periods within each cycle are generally much longer than the interglacial periods, which may be as short as 10,000 to 20,000 years. These cycles appear to be controlled by subtle rhythmic changes in the shape of Earth's orbit around the Sun (eccentricity), the tilt of the Earth on its axis (obliquity) and the 'wobble' of the Earth on its axis (precession). These changes can affect the amount of solar radiation the Earth receives, and more importantly when and where during each annual orbit the radiation falls (Fig. 5).

For example, the shape of the Earth's orbit around the Sun changes from near circular to an ellipse over a period of roughly 100,000 years. While this change in the eccentricity of the Earth's orbit has only a slight effect on the total amount of solar radiation received by the Earth each year, it does have signifi-

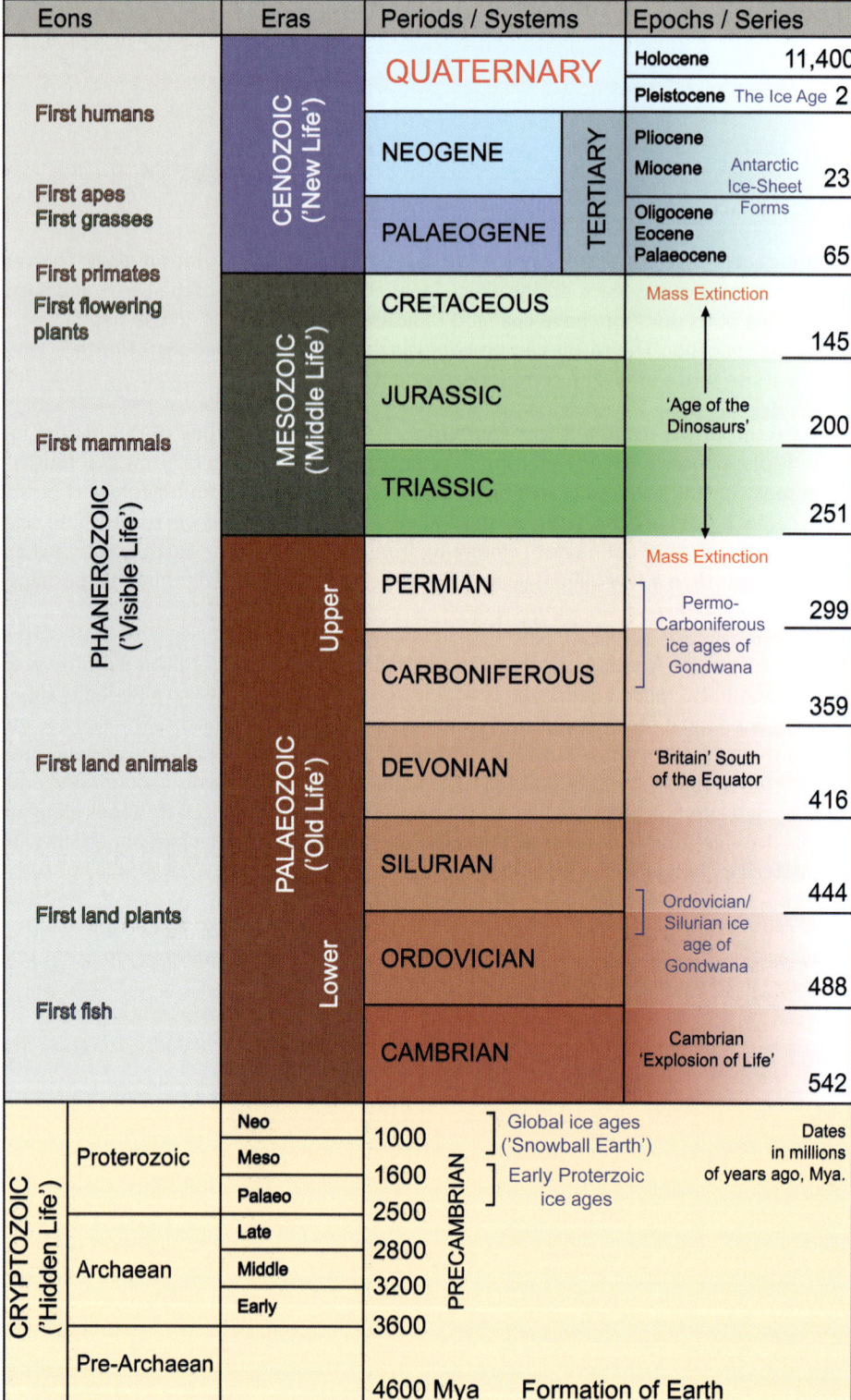

Fig. 4: The place of the Quaternary Period in a geological time scale. Note that the length of the Quaternary is greatly exaggerated in this chart. The dates of some earlier ice ages are also marked, as well as major landmarks in the evolution of life.

Fig. 5: Three types of variation in the Earth's orbit around the Sun affect the amount of solar radiation reaching different parts of the Earth's surface

cant seasonal effects. The tilt of the Earth's axis towards the Sun varies by a little under three degrees over a period of around 41,000 years. This again affects seasonality – the greater the angle of tilt, the greater the difference between summer and winter.

Working out the combined effects of these three variables – eccentricity, obliquity and precession – on the amount of solar radiation received by a particular latitude on the surface of the Earth at any given time in the past or future is mathematically very complicated. The man generally credited with cracking the problem, in the 1920s, was the Serbian mathematician Milutin Milankovitch. The long-term cycles are called Milankovitch cycles in his honour.

Milankovitch's astronomical theory was out of favour with geologists for much of the 20th century because it did not agree with the dates then given to the various glaciations recognised in the geological record. We now know that these dates were wrong. The match between Milankovitch cycles (a theoretical cause of the cycles of glaciation) and the oxygen isotope record from the sea bed (an actual, if indirect, record of the cycles of glaciation – see below) is much better. So Milankovitch's theory is back in favour. For example, the major glacial-interglacial-glacial cycles of around 100,000 years' duration that appear to have provided the underlying rhythm to the Earth's climate history over the past million years correspond, if imperfectly, to the major astronomical cycle of the changing eccentricity of the Earth's orbit around the Sun.

However, the history of the Earth's climate is much more complicated than can be explained by the astronomical theory alone. This is because the Earth itself and life on it have changed through time. On a geological timescale one of the most important changes is 'continental drift', or plate tectonics as it is more properly called. The configuration of the continents over the surface of the Earth is an important factor in whether the relatively subtle cyclical changes in solar radiation falling on the Earth will trigger an ice age. (There have been other ice ages in the past before the current Quaternary Ice Age, the last one around 300 million years ago – see Fig. 4). At present the position of the continent of Antarctica over the South Pole is conducive to an ice age – huge ice sheets have, of course, accumulated there.

A second major factor is the state of life on Earth. Due to concerns about global warming, the idea is now familiar that the composition of the atmosphere, particularly the concentration of greenhouse gases such as CO_2 within it, plays a fundamental role in our climate. Also familiar is the idea that life on

Earth – everything from bacteria to humans to rainforests – profoundly shapes the composition of the atmosphere due to the chemical processes involved in life and death.

Just by taking into account the three factors already mentioned – astronomical cycles, continental drift and life – it can already be seen how complicated the causes of glaciations must be. And there are many more, for example the pattern of ocean currents: were it not for the Gulf Stream, Britain would be much colder than it is today.

One important factor is that of 'positive feedback', which is anything that tends to amplify a change once it has been initiated. Snow and ice, being white, reflect heat away from the Earth. The more snow and ice there is over the surface of the Earth, the more heat is reflected away and the colder it gets. Because it is colder the area of permanent snow and ice expands (less of it melts). This results in it getting colder still, more snow and ice, and so on. The process will continue until something checks it.

So the causes of ice ages in general and glaciations in particular are many, often subtle and interconnected in complex ways. It is little wonder that the Earth's climate history – and future – is still so imperfectly understood. We will explore this further in a later section when we look at the evidence for rapid climate change events.

SOME OF THE MAJOR EFFECTS OF GLACIATIONS

Fig. 6: Britain as part of the European landmass. During the coldest parts of the Ice Age sea levels were a hundred metres or more below present levels. This map shows the extent of the British peninsula with sea levels at the –100m contour.

During glacial periods ice sheets developed and expanded, at times covering much of the Northern Hemisphere. As they grew in size, sea levels fell as much of the planet's water became locked up in the ice sheets. As a consequence, during the colder periods Britain was not an island but a peninsula of mainland Europe (Fig. 6). As the climate warmed up and the ice sheets melted, sea levels rose and Britain was once again, sometimes, an island. This sequence of events has occurred repeatedly in response to the climate cycles of the Ice Age. However, due to unique historical factors, no two events are precisely the same.

For example, at one time Britain was connected to continental Europe by a high ridge of chalk rock connecting Kent to the Artois region of France across what is now the Dover Strait. This ridge was exposed even at times of high sea level. During glaciations, however, a huge freshwater lake built up to the south of the ice sheet, in the area of what is now the southern North Sea. This lake was dammed on its southern edge by the chalk ridge. At some point – geologists do not yet agree when – this ridge was catastrophically breached. Flood waters poured through, creating a vertical-sided valley more than 10 kilometres wide and 50 metres deep. This was probably one of the most powerful flood events in the history of the Earth. It was only after the breaching of the chalk ridge that Britain would become fully an island during interglacial periods.

The story of the Ice Age is not, therefore, just one of slow processes unfolding at a 'glacial pace' over thousands of years. It is also punctuated by catastrophic events.

Quaternary climate change is responsible for much of the character of the British landscape. Glaciers, moving with huge force, shaped and reshaped the mountains in the highland areas of the west and north. They also carried huge quantities of material – crushed rock and other debris – into the lowland areas of the south and east. The courses followed by Britain's major rivers were determined by these changes. Modern patterns of soils and vegetation can only be understood in the light of Ice Age history.

THE DEEP SEA RECORD AND MARINE OXYGEN ISOTOPE STAGES

The pattern of Quaternary climate change is revealed in deep-sea sediments, through what is known as the marine *oxygen isotope record*. This record is inferred from the shells of tiny marine animals (foraminifera – 'forams' for short), the chemical composition of which reflects the chemistry of the ocean during their lives (Fig. 7).

The proportions of two different oxygen isotopes (oxygen atoms of different weight) in the sea water changes as ice sheets grow and then melt. The lighter isotope (^{16}O) is preferentially included in water evaporated from the surface of the sea, some of which is ultimately precipitated as snow. At times when the glaciers are large, more of this isotope is locked up in the ice sheets and it is consequently depleted in the sea water – the ratio of the lighter isotope (^{16}O) to the heavier isotope (^{18}O) is reduced. At times when the glaciers are small, as today, the opposite occurs. These changes are reflected in the composition of the foraminifera shells, which are formed using oxygen from the sea water.

Scientists can measure the oxygen isotope ratios in the shells and thus get an indirect (or proxy) indicator of the size of the glaciers, and hence of global climate, at the time the animals died. By drilling into the thick deposits of sediment containing the shells that have built up on the sea bed gradually over thousands of years, scientists can produce a continuous record of past climate change (Fig. 8).

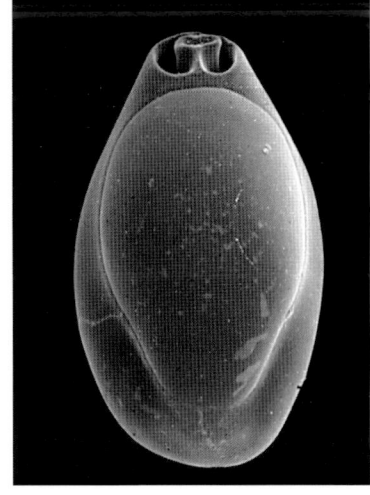

Fig. 7: A bed of foraminifera and a foram magnified 2000x

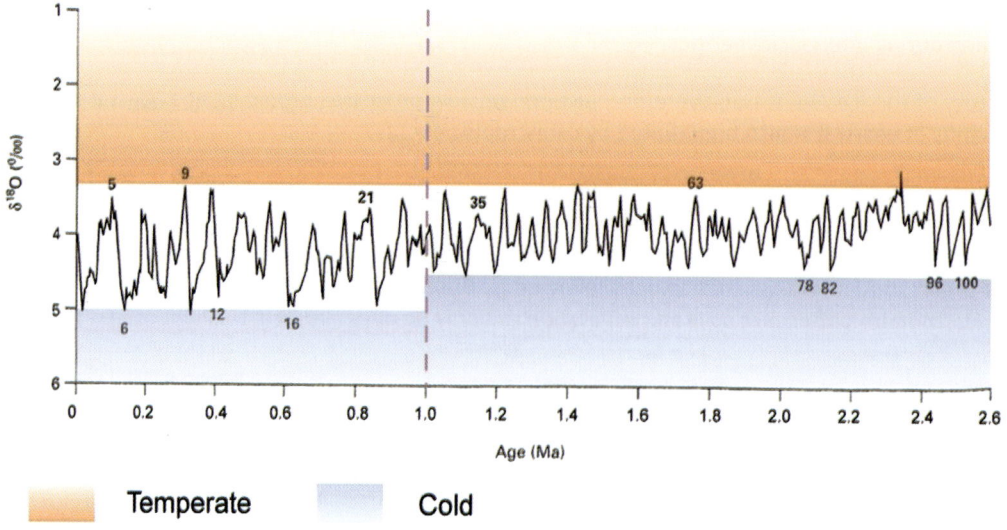

Fig. 8: Marine isotope curve recording climate over the last 2.6 million years. Some of the marine isotope stages have been numbered.

Each cold or warm period has been given a number, the *Marine Oxygen Isotope Stage* (MIS). Warm stages are given odd numbers and cold stages even numbers. The sequence starts with the present-day warm stage, Stage 1 (MIS 1). The most recent glacial stage, centring around 22,000 BP, is Stage 2 (MIS 2) and so on backwards (see Fig. 8). Most Quaternary scientists and Palaeolithic archaeologists now use the marine oxygen isotope record as the basic framework for their studies.

Ice-Core Records and Abrupt Climate Change

In the past couple of decades our understanding of climate change during the Quaternary has been transformed as a result of cores drilled not into the sea bed but into the ice sheets covering Greenland and Antarctica. These ice sheets are miles thick and drilling into them to obtain information on the history of climate and atmospheric change has been a major scientific, and logistical, triumph.

The principles are simple if the practice is difficult. Each year of snowfall compacts to form a new layer of ice over the ice sheet. It is possible to drill through the ice sheet and count each annual layer of ice in the cores recovered, back over hundreds of thousands of years. Each layer of ice contains within it information on the climate and atmosphere at the time it was laid down. The oxygen isotope ratios of the snow can be used to infer temperature change following the same general principles as those used to infer climate change from deep sea cores. A difference is that the isotope ratios are measured directly from the ice.

One of the most important things that the ice cores provide is much greater chronological resolution than the deep-sea cores. In principle one can measure changes year by year, although in practice there are all sorts of difficulties, especially as one goes further back in time. The results have been dramatic, as can be seen from a temperature curve for the past 100,000 years reconstructed from oxygen isotope measurements along the GISP2 core from central Greenland (Fig. 9). Instead of the gradual change traditionally associated with the Ice Age, it can be seen that the climate changed frequently, rapidly and violently. Some climate change was so abrupt that it would have been experienced within the course of

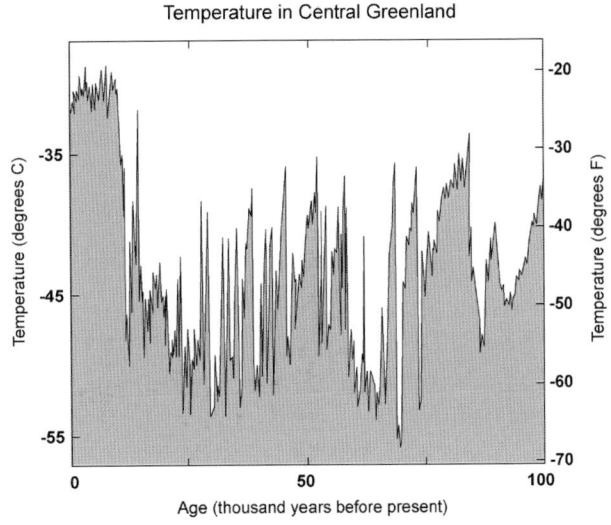

Fig. 9: A profile of temperatures in central Greenland during the past 100,000 years, reconstructed from oxygen isotope measurements along the GISP2 ice core

a human lifetime. For example, the data suggest that the transition from the last glacial period to the present interglacial (the Holocene) occurred in only 40 years. This has major implications for our understanding of how humans survived and adapted during the Ice Age, and indeed for the whole process of human evolution. Furthermore, it presents scientists with enormous challenges for understanding the causes and consequences of climate change, some of which may have been triggered by catastrophic events similar in nature to the breaching of the Channel ridge. Clearly, although Milankovitch cycles may provide the background 'pacemaker' to climate change, they are only part of the story.

Stable isotope analysis is only one of the techniques that can be used on ice cores to infer climate and atmospheric change. The ice layers also trapped air bubbles, dust and volcanic aerosols. From the bubbles of trapped air the quantities of greenhouse gases such as CO_2 and methane in the atmosphere can be measured. The effect of atmospheric pollution over the last 200 years is clearly registered in the most recent ice layers. Dust provides important information on prevailing winds and the storminess of the atmosphere, as well as on the general nature of vegetation cover. Large areas of poorly vegetated land – deserts – give rise to large quantities of dust in the atmosphere. Volcanic eruptions can be traced from their fall out trapped in the ice layers.

THE TERRESTRIAL RECORD AND ITS TERMINOLOGY

So far we have focused on the climate record that can be obtained from deep-sea cores and ice cores. Since the 1970s and 1990s respectively, these new scientific techniques have revolutionised our understanding of the nature and pattern of climate change during the Quaternary. They have provided us with a continuous, well-dated and detailed record. Quite rightly, they have attracted a great deal of attention (and funding). It is important to appreciate, however, that they provide us with only part of the story. If we want to reconstruct environments locally and regionally then we must turn to the terrestrial record. It is only through the land-based record that we can discover the effects that climate change had on the landscape, through the action of glaciers and rivers, for example. It is only from this record that we can learn what plants and animals were present, what the pattern of human occupation was, and how plants, animals and humans responded to climate change.

In contrast to the study of deep-sea and ice cores, the study of the terrestrial record has a history going back 150 years. Since the general acceptance by scientists in the 1840s of the evidence for an Ice Age, and the subsequent realisation that there had been not just one glaciation but several, a fundamental task for researchers has been to sort out the basic sequence of events. Glacial and interglacial episodes leave different kinds of sedimentary and other evidence on land, and through the principles of stratigraphy a sequence can be built up.

Scientists working in different countries have come up with different schemes. The practice has generally been to give each major glacial episode a name based on the place or area where the evidence was first found or is most characteristically represented. Thus the classic scheme relating to the Alpine area of Europe recognised four major glaciations – Günz, Mindel, Riss and Würm – each named after a river valley. The most recent of these glaciations, the Würmian, is called the Weichselian in north-western Europe, the Devensian in Britain and the Wisconsinian in North America. The potential for confusion is considerable, although these local schemes do have the merit of reflecting the fact that global climate change has different local effects. Obviously, the lower latitudes did not experience glaciations during 'glacial' periods. Here the effects of global climate change were different, with 'glacial' periods generally drier – resulting in shrinkage of the tropical rain forests and expansion of deserts – and 'interglacial' periods generally wetter.

Interglacial periods have also been given local names. For example in Britain a major interglacial period, the Hoxnian (now dated to around 400,000 BP), is named after a brickearth pit at Hoxne, in Suffolk, famous for being the first Palaeolithic site to be recognised in Britain if not the world. It is preceded by the Anglian Glaciation (around 450,000 BP and corresponding to MIS 12), the most extensive glaciation in Britain for which we have good evidence. It is named after East Anglia in general, where the sediments deposited during this episode are particularly well preserved. The most recent interglacial before the present one is called the Ipswichian (about 125,000 BP) and so forth.

Sorting out these different national and regional schemes, assigning particular exposures of deposits to one or other of the named episodes and correlating the different schemes one with another has preoccupied – and continues to preoccupy – Quaternary geologists. However, the deep-sea oxygen isotope record now provides a continuous record of global climate change that is independent of the terrestrial record, and most Quaternary scientists and Palaeolithic archaeologists have adopted this as the 'master framework' within which to work. Nevertheless, this does not solve all the problems. It is by no means straightforward to correlate the terrestrial record – say the sedimentary evidence for a particular climatic event – with the marine record. To do so requires, above all else, accurate and precise methods of dating.

Further complication is introduced by the evidence for abrupt climate change, operating on a scale of tens and hundreds rather than thousands of years, revealed by the ice cores. How are such events to be recognised in the terrestrial record, especially given that the dating techniques available are too imprecise to deal with such short timescales? These are some of the challenges Quaternary scientists now face.

The overwhelming challenge of the terrestrial record is, however, its incompleteness. Each successive glacial episode tends to erase much of the evidence for previous episodes. As ice sheets move over the landscape, they bulldoze and destroy much of the underlying sediment, altering river drainages and creating new landscape features such as glacial valleys and moraines.

A striking example of these processes can be seen in the history of the now obliterated Bytham River. Around 700,000 BP the Bytham flowed through the Midlands and drained into the North Sea through East Anglia. It was a very major river and appears to have been a principal route followed by the earliest humans to occupy Britain. Yet when the Anglian ice sheet surged over the British landscape about 450,000 BP, the course of the river was completely destroyed by the glaciers (Fig. 10). The only evidence we now have for the existence of this river is in the remnant sands and gravels deposited along its course, exposed as a result of quarrying. Indeed, it was commercial quarrying in the 1980s that first led to the discovery of this ancient lost river.

The challenges that face Quaternary scientists and archaeologists trying to understand the British terrestrial sequence are therefore many. It is an exercise of fundamental importance, however, because establishing the sequence and its local variations is the essential first step towards an understanding of the environmental impact of climate change. This then enables deeper questions to be addressed, such as how animals and early humans adapted to the changes, and the patterns of their occupation and evolution.

As with the case of the Bytham River, it is commercial quarrying that produces much of the evidence, in the form of exposing important but often deeply-buried deposits during extraction. These deposits may contain animal bones, plant and insect remains, snail shells and stone tools (see chapters 4-6). Although often difficult to recognise to the untutored eye, these deposits are of critical importance to reconstructing past environments and revealing the pattern of past human occupation. There is also an urgent need to date these deposits using the latest scientific techniques (chapter 7), so their place in the sequence can be established.

BRITISH QUATERNARY STRATIGRAPHY

The main later stages of the British Quaternary sequence as they are now understood are summarised in Figure 11. The principal glacial and interglacial periods are named together with the corresponding

Fig. 10: The great Anglian ice sheet of around 450,000 years ago obliterated the Bytham River and diverted the Thames into a new course. Note the massive proglacial lake that has formed just to the south of the ice sheet, in what is now the southern North Sea basin.

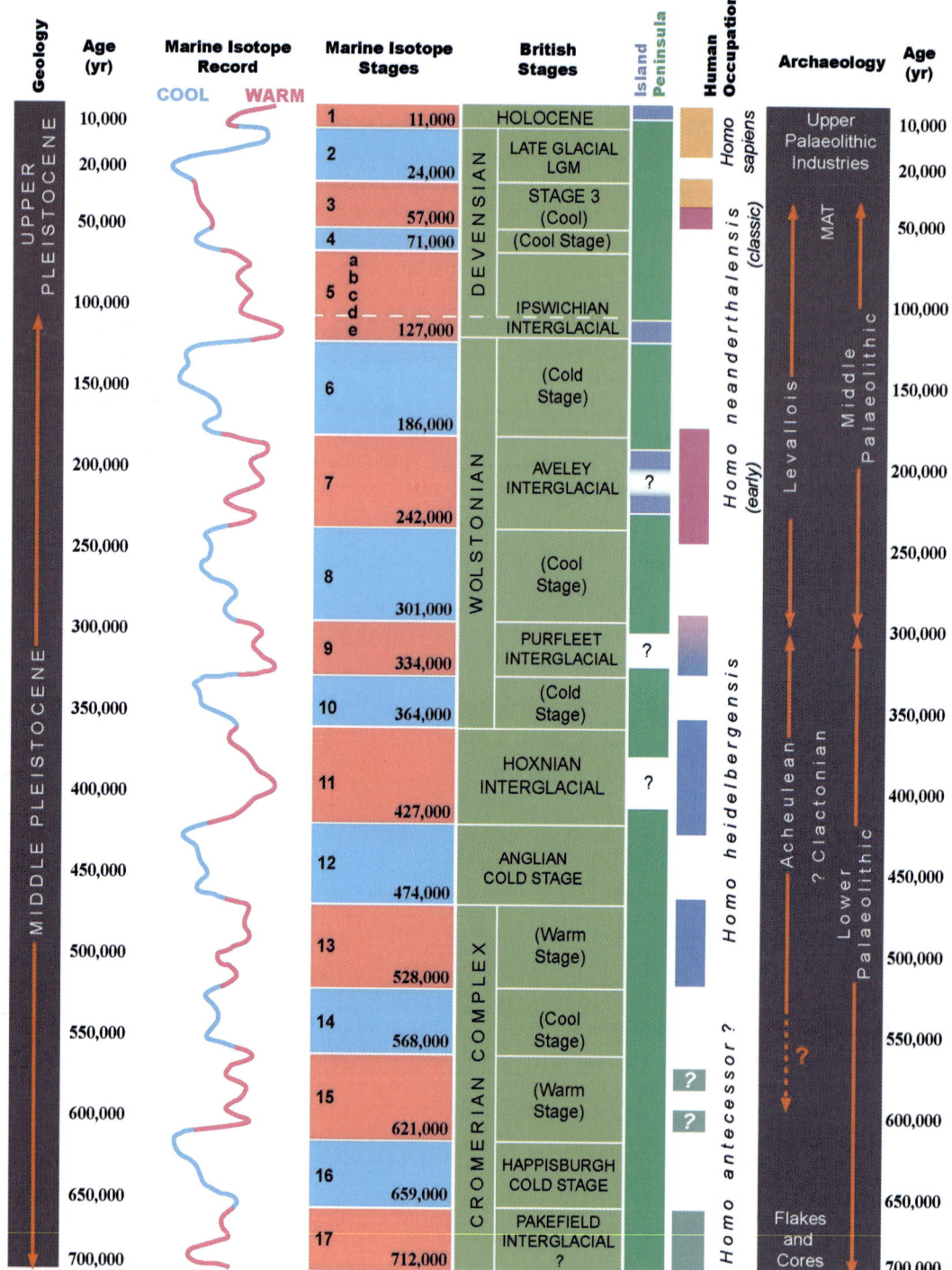

Fig. 11: A chart showing the later stages of the Quaternary sequence in Britain (Middle and Upper Pleistocene) as understood today. The chart shows how the saw-tooth curve of the marine isotope record is divided up into numbered marine isotope stages (the divisions between the stages and the dates assigned to these are necessarily somewhat arbitrary). The chart also attempts to relate the marine isotope stages to the terrestrial record, with its traditional named stages. Further columns relate the climatological and geological sequences to the pattern of human occupation and the archaeological sequence, discussed in chapters 3 and 6 ('MAT' - Mousterian of the Acheulean Tradition; 'LGM' - Last Glacial Maximum).

marine isotope stages. It will be noted that the oxygen isotope record suggests that some of the main glacial and interglacial periods recognised in the terrestrial record (for example the Devensian and the Cromerian) are in fact rather complex and encompass several isotope stages.

Much research remains to be done. Quarries are a vital source of new information that can help fill many of the gaps in our knowledge as they expose Quaternary sediments that would otherwise remain buried beneath our feet.

QUATERNARY ENVIRONMENTS

Not surprisingly, the dramatic climate changes of the Quaternary have significantly influenced the animals and plants that lived in Britain. As the climate has changed, organisms have migrated, evolved or become extinct. Therefore, through time, the landscapes and ecologies of the British Isles have been constantly changing in character and composition.

Every animal and plant species has a particular set of ecological requirements that need to be met in order for it to survive and reproduce. These are ultimately controlled by climate. Therefore as the climate deteriorated and ice sheets expanded, the animals and plants that required warmer conditions retreated southwards and eastwards. Correspondingly, cold-loving species expanded their range southwards. Each found newly available niches. As the ice sheets melted, the process went into reverse. Warm-loving plants and animals emerged from their refugia in the south and re-colonised the north again; cold-loving species retreated.

Fig. 12: Glacial environment: a multi-channelled ('braided') river flowing through a tundra landscape

However, it was not a simple pattern of ebb and flow. Unique historical factors came into play. As we have seen, no two glacials or interglacials were precisely the same. The landscapes of each were often very different. Appropriate niches may not reappear or may be occupied by competitors. Different species migrate and colonise by different methods and at different speeds, giving rise to a characteristic succession of species. However, for some species migration routes may become cut off by rising sea levels or other landscape change. Furthermore, climate change during the Ice Age drove not only the movement of plants and animals – the most basic response – but also for some species (notably mammals, including humans) their evolution and extinction. Consequently, the animals and plants of Britain have varied through the Ice Age.

This variation can be put to practical use for dating, through a technique known as biostratigraphy (see also chapter 7). Each major period of the British Ice Age is characterised by a unique suite of animals. Defining what this suite of animals is for each period enables newly-discovered deposits with appropriate fossil remains to be fitted in to the overall sequence. Some of the most useful animals for this purpose are small ones, such as different species of voles. Thus even apparently insignificant remains can be of great importance.

Fig. 13: Interglacial environment: deciduous woodland

During cold periods, those parts of the British landscape not covered by ice were characterised by tundra-like habitats, colonised by sparse vegetation such as grasses and mosses, similar to Alaska or Siberia today (Fig. 12). The fauna was very different, however, with cold-adapted, now extinct, Ice Age animals such as mammoth and woolly rhino roaming tundra landscapes criss-crossed by braided rivers fed by glacial melt waters.

Interglacial environments looked very different. The open landscapes were gradually replaced by dense deciduous woodlands (Fig. 13) inhabited by warm-adapted animals such as fallow deer. More open areas could be found along river valleys and coasts, and on upland landscapes such as Salisbury Plain and the North Downs. The well-drained, fertile soils that formed on these chalk lands (Fig. 14) supported rich grasslands that provided grazing for large herds of herbivores. The range of animals found in Britain during many interglacials might seem surprising and more the sort of thing that would be expected on the East African savannah, including lions and spotted hyaenas. Straight-tusked elephant and

Fig. 14: Interglacial environment: chalk downland

a couple of species of extinct rhinoceros were present during most interglacials, but hippopotamus and macaque monkey were also present during some. Other animals, such as wild horse, bison and brown bear are tolerant of a range of environments and can be found in both cold (but not extreme) and warm periods.

It is misleading, however, to think of the climate of Britain as just either cold (glacial) or warm (interglacial). We have already noted that the ice-core record indicates that there was often quite radical fluctuation of climate within both glacial and interglacial periods. Furthermore, the transitions from an interglacial period to a glacial period and from a glacial period to an interglacial period were times when the climate was neither cold nor warm but cool. Thus the environments of the Ice Age were many and varied.

Archaeological evidence suggests that Britain was abandoned by humans during the coldest parts of glacials, when those parts of the country that were not covered by ice sheets were an arctic wasteland. They were present during most interglacials and also during the cool periods at the beginning and end of glacials. One interglacial during which humans appear to have been absent is the last interglacial, the Ipswichian, about 125,000 BP, when temperatures were four degrees warmer than they are today. This may be largely explained by Britain having become an island. The species of human present in Europe at the time – Neanderthals – had likely retreated south and east during the previous cold stage and were unable to recolonise in time before Britain was cut off by rising sea levels.

THE RECORD FROM RIVER VALLEYS

Significant remains from the Ice Age can be found in a very wide range of situations, from caves to the bed of the North Sea. However, the single most important context in which these remains are found is river valleys, where they may be found above, within and beneath the deposits of sands and gravels laid down by the rivers. This includes the valleys of extant rivers, such as the Thames, extinct rivers such as the Bytham mentioned above, and extant rivers that have significantly altered their course, such as the Medway or the Thames. In a nutshell, this is why sand and gravel quarrying is so important to the study of the Ice Age. It is therefore essential to understand in general terms how these river valleys formed and how the various sediments found within them (known as fluvial sediments and including of course sands and gravels) came to be laid down.

The predominant factor in determining the course of rivers is obviously topography – rivers 'find their way' down slope. As the overall structure of the topography of southern Britain comprises highlands in the west and lowlands in the east, then the predominant pattern for major rivers, such as the Thames and the ancient Bytham, is to flow from west to east and drain into the North Sea. At times when sea level was lower and much of the North Sea was dry land the journey of the Thames and other rivers to the sea would have been longer. After the English Channel had formed through the catastrophic breaching of the chalk ridge across the Dover Strait (see above) the Thames would have flowed into the mighty 'Channel River', flowing south-westwards, and also joined from the east by the Rhine, the Meuse and the Somme.

Glaciers affected rivers in four main ways. They were a major physical element of the topography, diverting some rivers and obliterating others. Their sheer weight depressed the Earth's crust beneath them, lowering the land significantly. (After they had melted and the weight was released the land began to rebound – Scotland and Scandinavia are still rising today following the melting of the ice sheets of the last glaciation.) Acting in the opposite direction, the presence of glaciers worldwide locked up water and led to a global lowering of sea levels so, as has been mentioned, rivers had further to travel. Finally, the melting of glaciers fed massive amounts of water, at high energy, into river systems.

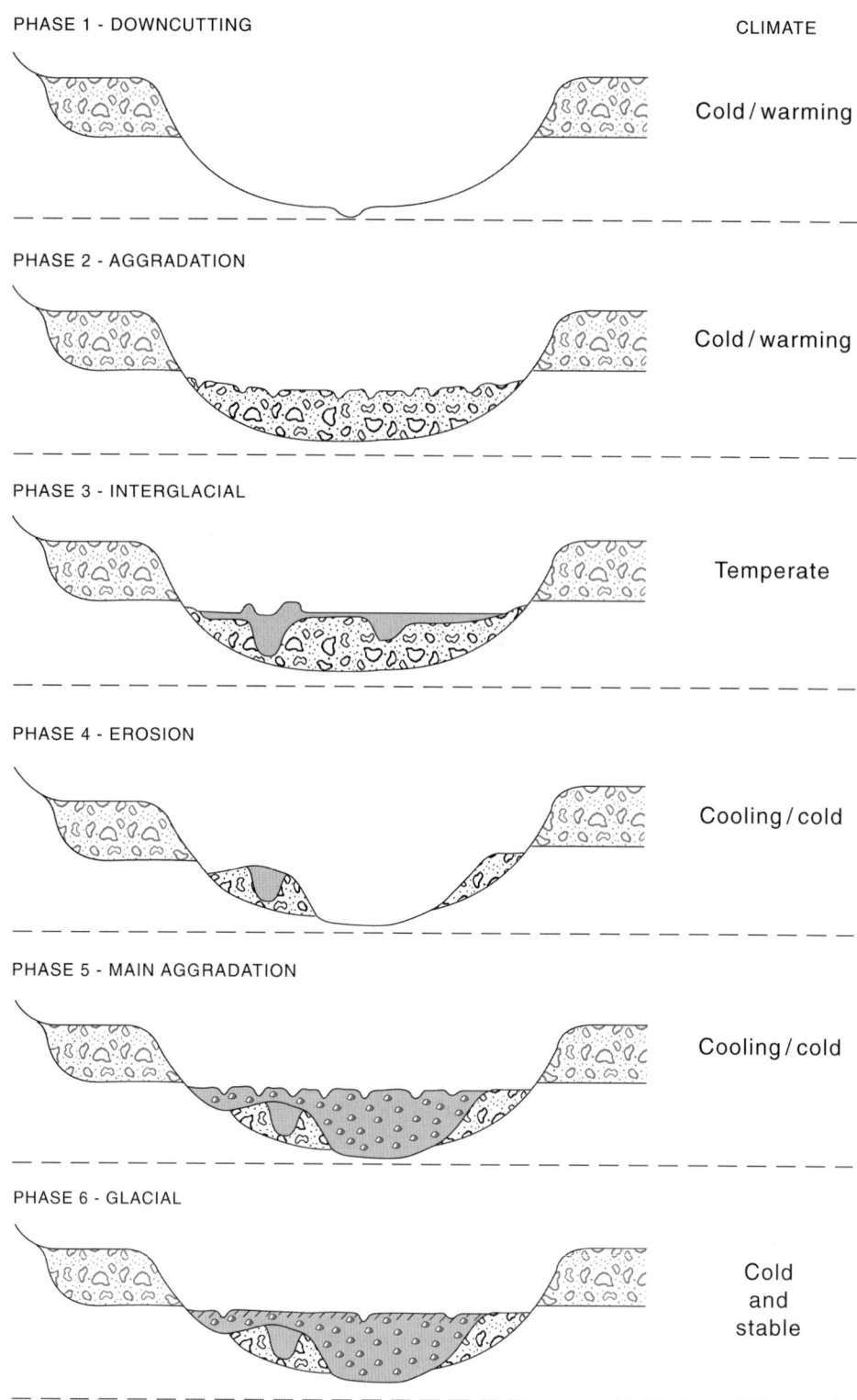

Fig. 15: Six-stage developmental model of river terrace formation according to David Bridgland. See main text for explanation.

Where rivers have survived successive glacial–interglacial cycles – as has, for example, the Thames – the configuration of their valleys and the sedimentary deposits within them bear witness to this history. The most striking aspect of this is the formation of a sequence or 'staircase' of relatively flat river terraces down the sides of lowland river valleys. These are features that can sometimes be seen fairly easily today.

The terraces are surviving fragments of former floodplains of the river. Lowland river valleys and terrace staircases were generally formed by repeated processes of, on the one hand, *downcutting* by the river and, on the other hand, the accumulation of large quantities of fluvial sediment deposited by the river, a process known as *aggradation*. According to a model developed by the geomorphologist David Bridgland, the general sequence, at least for some rivers, appears to be as follows (Fig. 15).

PHASE 1 – DOWNCUTTING (COLD/WARMING)

As the climate begins to warm at the end of a glacial period, huge quantities of water are released from the melting snow and ice. This results in very high-energy rivers that cut down into their valleys and through the deposits of sands, gravels and other alluvium that had been laid down by the river in an earlier phase. Surviving remnants of these deposits form a terrace, now 'high and dry' above the river that has cut deeper down into the valley.

PHASE 2 – AGGRADATION (COLD/WARMING)

Although the climate is warming, temperatures are still very cold. The high-energy rivers flow over a bleak landscape consisting of wide expanses of gravel colonised by occasional hardy tundra plants and mosses. The rivers – known as braided rivers – are very wide and have multiple channels separated by islands of eroded gravels. As well as downcutting, these rivers also carry large quantities of coarse fluvial sediment which they deposit in the lowered river valley (aggradation).

PHASE 3 – INTERGLACIAL (TEMPERATE)

As the climate enters the full interglacial stage, the rivers become much quieter, generally flowing across the valley floor in single meandering channels. Both because they are much less powerful and because plant growth has stabilised the valley sides, these temperate rivers transport much less coarse sediment. Instead they carry a combination of silts and fine sands, usually laid down in abandoned meanders of the river known as oxbow lakes. These fine-grained deposits are only rarely preserved but when they are they are very valuable because they often contain fossils and organic remains.

PHASES 4 AND 5 – EROSION AND MAIN AGGRADATION (COOLING/COLD)

The next stage in the process occurs as the temperature starts to cool, heading for the next glaciation. There is both erosion (Phase 4), but not enough for terrace formation, and major aggradation (Phase 5) as declining vegetation liberates sediment from the valley sides.

PHASE 6 – GLACIAL (COLD AND STABLE)

Finally, as the climate enters the full glacial stage, the rivers become stable again. This is because a lot of the potential discharge is locked up in permafrost. Then, as the climate begins to warm towards the next interglacial, the cycle starts over again.

Each cycle of downcutting, terrace formation and aggradation thus occupies a whole glacial-interglacial cycle. When the cycle is repeated several times the result is a terrace staircase, the highest terrace being the oldest, and each successive terrace below that being progressively younger down to the present-day river. Furthermore, it was noted in an earlier section that since about a million years

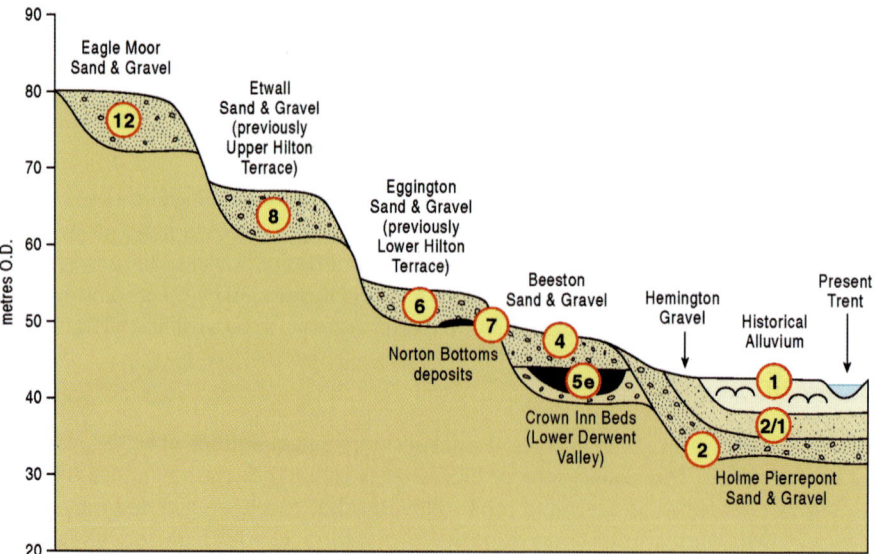

Fig. 16: An idealised section through the Middle Trent terrace 'staircase'. The circled numbers indicate the marine isotope stage to which the deposits have been assigned.

ago the duration of each major glacial-interglacial cycle has been around 100,000 years (apparently corresponding to the Milankovitch cycle driven by the changing eccentricity of the Earth's orbit). This means that, in theory, each terrace should represent a record of around 100,000 years of fluvial activity. Also, in an ideal world, each terrace would contain within its make up material deposited during a

Fig. 17: The Whitemoor Haye woolly rhinoceros (*Coelodonta antiquitatis*) with its discoverer, machine driver Ray Davies. This partial skeleton was found in 2002 in MIS 3 deposits at Whitemoor Haye quarry, on the confluence of the Trent and the Tame.

warm (odd numbered) marine isotope stage and, above that, material deposited during the cold (even numbered) marine isotope stage that followed it. The latter material will make up the vast bulk of the terrace and will largely comprise coarse sands and gravels. According to the model sketched above, most of the cold-period material will actually have accumulated towards the end of the glacial, at the transition to the next interglacial.

If all this held true one could have the sensation, climbing up the sides of a river valley and crossing the steps of each terrace, that with each step reached one had gone back 100,000 years in time. Of course, this is just a hypothetical model, a useful approximation, not expected to be precisely right in every or indeed any real-world case. The real world is much more messy and complicated. Nevertheless, the model provides a very useful way to think about river terrace formation and is often not that wide of the mark.

A real-world example is given in Figure 16. It shows an idealised section through the Middle Trent terrace 'staircase'. There are five terrace steps and each terrace is numbered with the marine isotope stage to which it has been assigned. The bottom step, the floodplain terrace over which the present-day Trent flows, contains sands and gravels that were laid down towards the end of the last glacial period (MIS 2) and the beginning of the current interglacial (MIS 1), that is between about 10,000–25,000 BP. No deposits dating to MIS 3 (about 60,000–25,000 BP) are shown, but these have been found under equivalent gravels at Whitemoor Haye quarry at the Trent-Tame confluence. Here the deposits contained very well preserved plant, beetle and mammal bone assemblages, including the partial skeleton of a woolly rhinoceros (Fig. 17). (Although MIS 3 is an odd-numbered and therefore 'warm' marine isotope stage, for much of the stage the climate was, in fact, rather cold but not glacial, explaining the presence of the cold-adapted animals such as the woolly rhino.)

The step up to the next terrace takes us back around 100,000 years. The Crown Inn Beds were laid down in the last interglacial, the Ipswichian (MIS 5e), around 125,000 BP. At the end of the 19th century they yielded plant material and large animal remains, mostly hippopotamus, which is a characteristic indicator of this interglacial. The overlying Beeston Sand & Gravel was laid down in the cold-climate conditions of MIS 4 around 80,000 years ago.

Stepping up to the next terrace, the Norton Bottoms deposits date to the penultimate interglacial (MIS 7), centred around 210,000 BP and thus just a bit less than 100,000 years earlier than the Ipswichian. The overlying Eggington Sand & Gravel accumulated during MIS 6, between about 186,000 and 127,000 BP. And so on up to the top of the staircase, although the marine isotope stage assignments become less certain in the higher, older parts of the staircase.

The highest, earliest terrace, the Eagle Moor Sand & Gravel, is assigned to MIS 12, which equates with the great Anglian glaciation of around 450,000 BP, when the ice sheets reached their most southerly extent in Britain. The sequence has to stop there because any predecessor to the River Trent would have been completely obliterated by this glaciation. In accordance with the model sketched above, the Eagle Moor Sand & Gravel would have been laid down towards the end of the Anglian, when the Trent river system first became established. It will be noticed that MIS 10 is missing from the Trent sequence, and it is possible that the uppermost terrace should be assigned to this stage rather than MIS 12.

It should be apparent by now that lowland river valleys are in essence 'archives' of the Ice Age. In theory, a staircase of gravel terraces has the potential to provide a record of environmental change along a river valley going back hundreds of thousands of years. The study of these archives gives us the possibility to understand, on the ground, at the local and regional level, the environmental effects of global climate change. Put a number of these archives together, and we have the opportunity to explore patterns over wide geographical areas, with a time depth rivalling that of the ice cores. Furthermore, these archives contain not just information about past environments but also, because they sometimes contain archaeological remains such as stone tools, about patterns of human occupation and behaviour. As will be described in the following chapter, in the older, higher terraces this concerns human species very different from us.

The most frequent and important way in which these invaluable archives are opened is through commercial sand and gravel quarrying. This is why it is so important that scientists, archaeologists and the quarry industry work together to extract the evidence as it is turned up. Unfortunately, good evidence, for example deposits contained well-preserved assemblages of fossils enabling detailed environmental reconstruction, is comparatively rare. The archives are patchy and fragmentary. This makes it even more important that when good evidence does turn up it is recognised and scientifically studied; this handbook is aimed towards that end. Only very rarely, as we emphasised earlier, does this involve significant disruption to quarrying.

The main reason why the archives are so patchy is because of the way that they were formed. From the description we have given of the processes of downcutting and aggradation, and of the high-energy fluvial forces involved, it will be appreciated that the chances of delicate fossil material surviving are extremely slender. The chances of survival for larger, more durable material, such as elephant teeth and stone handaxes, are better. However, the problem here is that very little of this material will be found *in situ*. The river will have moved it, sometimes over considerable distances. Things found together may well not belong together, which makes interpretation difficult.

The most extreme version of this problem is when material is found in a geological deposit to which it does not belong at all. This is likely to be very common. As a river cuts down through a floodplain it carries away the deposit with its cargo of fossils and stone tools and redeposits it. Thus material is easily carried from a higher terrace to a lower one and may be 100,000 years or more older than it seems to be from the context in which it was found.

These problems are not insuperable. For example, study of the degree to which stone tools have been rolled, and the characteristic patterns of wear this produces on the artefact, provides some handle on the extent to which they are likely to have been moved from their original place of deposition.

The archives of the river valleys are therefore difficult to read and interpret, but the reward in terms of the insights into our past that they provide makes it more than worth the effort.

Chapter 3: Human Evolution and Ice Age Britain

In this chapter we provide some background on human evolution, emphasising the evidence from Britain and particularly that recovered from quarries. The genus to which we belong, *Homo*, and of which we *Homo sapiens* are now the sole surviving species, evolved around the same time as the start of the Ice Age. *Homo sapiens* itself evolved towards the end of the Ice Age. Thus the story of human evolution and the story of Quaternary climate and environmental change are intertwined strands of the same narrative.

Britain is not rich in the fossil remains of Pleistocene humans. To a certain extent this is not important (and we can always keep looking and hope). What is most distinct about humans is not what we look like but what we do. The main source of information on human behaviour before the invention of writing, which occurred long after the end of the Pleistocene, is archaeology. From archaeological remains we can infer something of how early humans behaved, and from that we can infer something of their abilities and limitations, and hence their mental capacities.

For much of the Ice Age, Britain was on the periphery of the inhabited world. This does not mean, however, that the British evidence is unimportant to the story of human evolution. Indeed, because of Britain's peripheral location, as well as the quality of its archaeological record and the long tradition of research, it can be used to study questions that are difficult to address as effectively anywhere else. For example, we can study how the humans of the Pleistocene – all of whom were foragers, surviving by hunting wild animals, gathering wild plants and perhaps scavenging carcasses – responded to climate and environmental change, and how they used the landscape and its resources. In Britain our ancestors and evolutionary cousins (e.g. the Neanderthals) were repeatedly stretched to their limit and beyond, as the periods of human absence testify. We begin our brief overview of human evolution with the divergence of the human line from that of the other surviving apes.

Genetic studies suggest that humans and chimpanzees (our closest living relative) shared a common ancestor, an ape-like creature, about 5–7 million years ago. Since that time, the evolution of chimpanzees and humans have taken divergent paths, with the species considered ancestral to modern humans being termed hominins.[1] Most of these hominin species followed different evolutionary paths which ultimately led, in all cases except our own, to extinction. Palaeoanthropologists (scientists who study ancient human fossils) have found many fossil remains of these early hominins in Africa, and until around two million years ago the story of human evolution appears to have been exclusively confined to Africa. Although the general outline of the story is clear, precisely which species belong on the direct path to modern humans and which belong to other evolutionary branches remains a matter of great controversy.

The principal characteristic of these early African hominins, up to around 2.6 million years ago, is that they evolved a peculiar mode of locomotion, 'bipedalism' or habitually walking on two legs rather than four. Other than this, they would have been more similar to the living apes of today than to modern humans (Fig. 18). Their brain size was closer to that of a chimpanzee than to a modern human and there is no evidence that they made even simple stone tools (although they would probably have used unmodified stones, stripped twigs and other items as tools to get at food like nuts and insects, just as chimps do today).

---

[1] Until recently, these species were known as hominids. Due to the reclassification of the family relationships between humans and apes, this term is now used to refer to both humans and other living apes, whilst modern humans and their direct ancestors are known as hominins.

Beginning around 2.6 million years ago (and thus at the start of the Quaternary period), these hominins started to make simple stone tools and to develop significantly bigger brains. The stone tools, mainly just choppers and sharp flakes struck from a pebble, would have been used for cutting meat from bones and smashing up bones to extract the nutritious marrow from within (Fig. 19). These changes signal the appearance of the new group of related species (a genus) known as *Homo* of which we, the species *Homo sapiens*, are part (Fig. 20). From this point onwards it is legitimate to refer to all these related species as humans, albeit different kinds of humans to ourselves. Only some of these human species (again precisely which ones is a matter of controversy) were directly on the line leading to us; it is better to refer to those species that were not on this line informally as 'cousins' rather than 'ancestors'. However students of the Ice Age are interested in all human species, irrespective of whether they were on the direct line to us or not. Part of the fascination and importance of the Ice Age lies in the opportunity it provides to study a range of human species. This enables us to put our own species into perspective.

The simple stone tools found from around 2.6 million years ago are the earliest manufactured objects of which we have any record. The archaeological record begins at this point. This earliest archaeological period is known as the Lower Palaeolithic (i.e. the first part of the Old Stone Age). It lasted until around 300,000 BP and thus, compared with all later periods of cultural history, is a vastly long period, more than two million years in duration.

Around 1.8 million years ago a new human species evolved in Africa, *Homo erectus*. This was perhaps the first human species to spread beyond Africa, and occupied large parts of Asia, with famous fossils having been found in China ('Peking Man') and Indonesia ('Java Man').[2] Their braincases were significantly larger than those of earlier hominins, reaching around 1,000 ml, or about 75 percent of the average values for modern humans today. However, the skulls were thick and low, with a flat forehead and a strong bar of bone running across the top of the eye sockets. The lower jaws were heavy and chinless. From the neck down *erectus* was similar to ourselves, although much more thick boned and heavily built. The robust nature of the skull and skeleton is indicative of a tough, physically challenging lifestyle (Fig. 21).

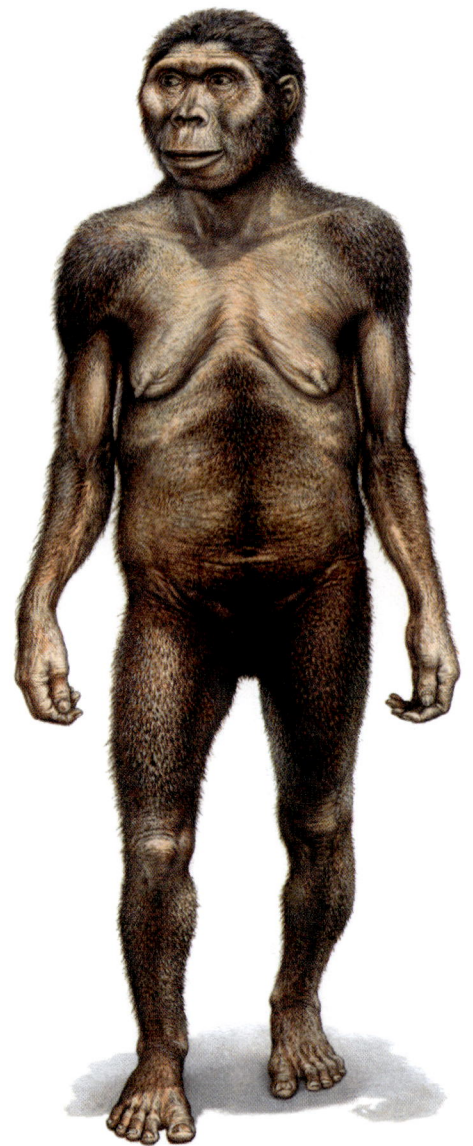

Fig. 18: An artist's reconstruction of 'Lucy', an early hominin from Africa. Lucy is a partial skeleton of the species *Australopithecus afarensis* ('southern ape from the Afar locality [of Ethiopia]') and dates to around 3.2 mya. Although bipedal, she was only just over 1m tall and had a brain size similar to that of a chimpanzee.

---

[2] Some scholars believe that the fossils attributed to *Homo erectus* represent two different but closely related species: *Homo ergaster* in Africa and its descendent, *Homo erectus*, in Asia.

Fig. 19: Early stone tools from Africa. Pebble tools discovered at the Olduvai Gorge, Tanzania, about 1.8 million years old.

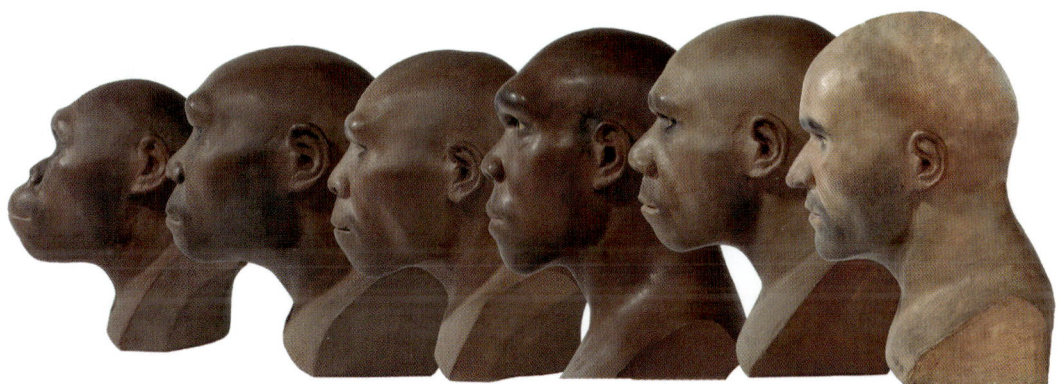

Fig. 20: From left to right: *Australopithecus, Homo erectus* (Java)*, Homo erectus* (China)*, Homo heidelbergensis, Homo neanderthalensis, Homo sapiens*

Around 1.6 million years ago a new type of stone tool, the handaxe, appeared in Africa and much later spread to Europe and western Asia (Fig. 22). The manufacture of handaxes involves considerably more skill than the production of simple flakes. Multiple-flake removals are necessary to make a handaxe, and they are frequently regular and symmetrical in shape. A 'mental template' of the desired end product is required in the mind of the maker, and handaxes imply a cognitive advance. (Handaxes are described in more detail in chapter 6.)

The earliest handaxes are associated with *Homo erectus*, although not all members of this species used them; some used simpler chopper tools instead. For example, handaxes did not spread to China. Handaxes were also used by human species later than *erectus*. Indeed this technology persisted sub-

Fig. 21: Reconstruction of a *Homo erectus* group, based on findings at Zoukoudian cave near Beijing ('Peking Man')

stantially unchanged for over a million years. This extremely slow rate of cultural change, with long periods of apparent stasis, is one of the principal characteristics of the Lower Palaeolithic. It indicates that these early humans were very different to us. It is unlikely that they possessed complex grammatical language like ours.

Figure 23 provides a diagram showing the possible evolutionary development of the human genus from *Homo erectus* onwards, and the spread of humans out from Africa. The earliest evidence we have for a human presence on the fringes of Europe comes from the site of Dmanisi, Georgia, in the Caucasus. Here, five small-brained human skulls dating to around 1.7 million years ago have been uncovered, associated with a very simple flake and cobble stone-tool technology. To which species these fossils should be assigned is much debated but they may be an early form of *erectus* or a distinct species which has been called *Homo georgicus*.

Fig. 22: A handaxe. This illustration is of an example found in the brickearth pit at Hoxne, Suffolk in 1797. It was one of the first examples to be correctly assigned, on the basis of the circumstances of its discovery, to "a very remote period indeed".

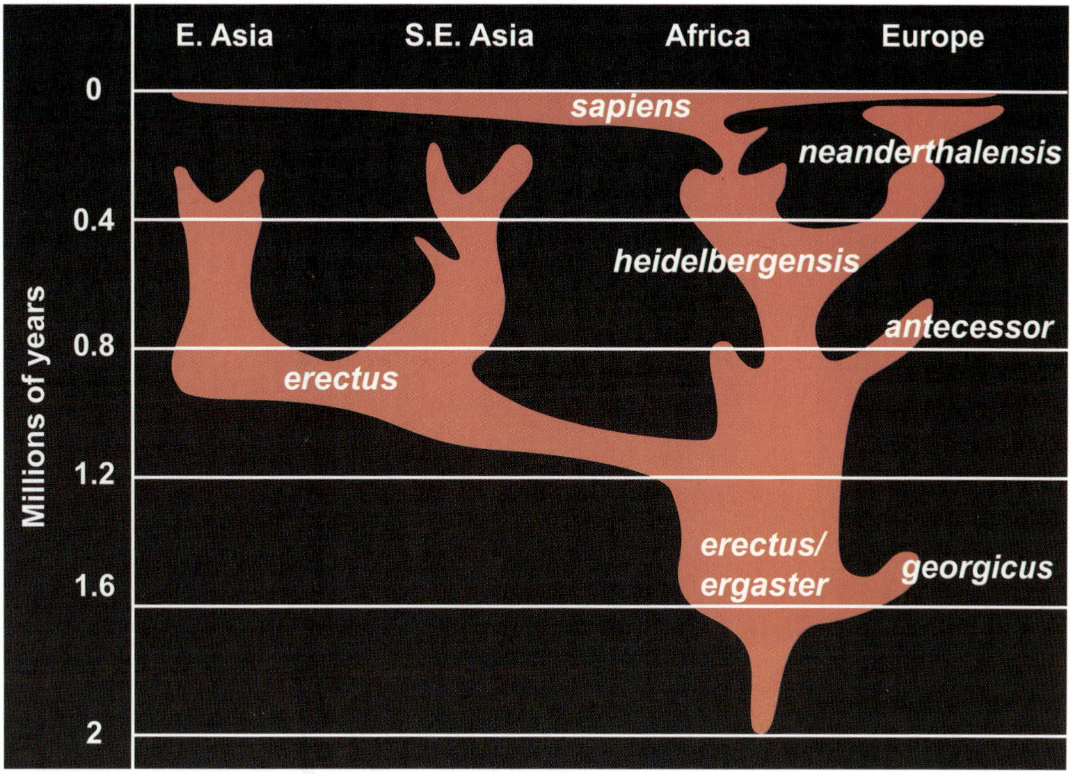

Fig. 23: An evolutionary 'tree' showing the possible development of the human genus and its expansion out of Africa. Note how the different species – E. Asian and S.E. Asian *Homo erectus, Homo georgicus*, *Homo antecessor*, *Homo neanderthalensis* and *Homo sapiens* – are all separate branches of the tree, rather than species following a linear succession up a branch-less 'trunk'.

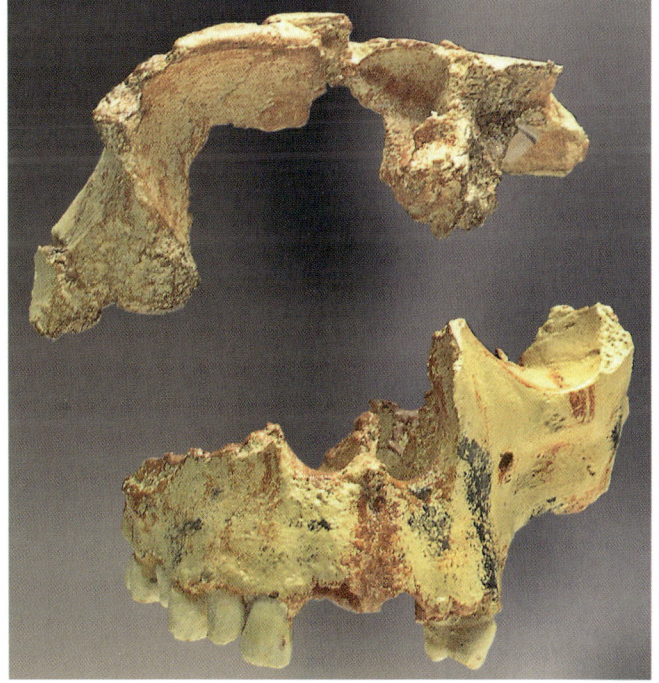

Fig. 24: Cast of a partial child's skull from Gran Dolina, Atapuerca in northern Spain. The skull is around 800,000 years old and has been assigned to the species *Homo antecessor*.

Fig. 25: Some of the humanly-struck flint flakes from Pakefield, Suffolk, dating to around 700,000 years ago

However, it is not until around 800,000 years ago that we have good evidence for the presence of humans in southern Europe, from the sites of Ceprano in Italy and, especially, Gran Dolina in Spain.[3] At Gran Dolina, in the Atapuerca Hills of northern Spain, numerous fragmentary human fossils belonging to several individuals, both adults and children, have been found (Fig. 24). Intriguingly, many of the bones are covered with cut marks made by stone tools, suggesting cannibalism. The tools in question are simple flakes and choppers; handaxes are absent. Once again, the species attribution is disputed. The Spanish researchers argue that the fossil remains are more 'modern' in many of their characteristics than *erectus*, and have proposed a new species, *Homo antecessor* ('Pioneer Man'), a descendent of *erectus*.

Until very recently it was widely believed that there was no good evidence for humans in northern Europe before around 500,000 BP. This changed with the publication in 2005 of evidence from Pakefield, Suffolk, for human occupation dated to around 700,000 BP.

The fossil locality at Pakefield is situated on the East Anglian coast under a cliff of glacial deposits exposed by the sea. Although quarrying is the most important mechanism for creating exposures of important deeply-buried Pleistocene deposits, the sea also plays an important role. Both then proceed to destroy the evidence they uncover unless action is taken to record it.

The deposits at Pakefield have for over a hundred years produced fossil bones of a wide range of animals. These include large herbivores such as an extinct species of hippopotamus, rhinoceros, straight-tusked elephant, bison and giant deer, as well as carnivores such as lion, a scimitar-toothed cat, hyaena and wolf. But evidence for human occupation long proved elusive. However, in recent years collectors reported what appeared to be simple but humanly-struck flint flakes from the locality. Although it could not be certain that these were definitely associated with the fossil-bearing deposits, this was confirmed by formal excavation under the auspices of the Ancient Human Occupation of Brit-

---

[3] New finds from the Sima del Elefante site in the Atapuerca Hills of northern Spain now provide evidence, including human fossil remains, for occupation even earlier, about 1.1–1.2 million years ago.

ain (AHOB) project. Over thirty worked flints in fresh condition have now been recovered, including several 'retouched' flakes (Fig. 25). Retouching (described more fully in chapter 6) involves removing small secondary flakes from a struck flake; it is a secure indicator of human manufacture.

From the huge wealth of fossil remains recovered from the Pakefield deposits – mammals large and small, fish, amphibians, beetles, molluscs and plants – together with the sedimentological evidence, it has been possible to reconstruct a very detailed picture of the climate and local environment at the time the deposits formed. The climate was Mediterranean, with warm, dry summers and cool, wet winters with temperatures that remained above freezing. The environment was that of an extensive river estuary, with marshy ground and pools, and oak woodland and open grassland further from the river. The estuary flowed into the western side of a huge bay, opening northwards into what is now the southern North Sea. The estuary probably relates to the river system of the lost Bytham River (Fig. 30, p.33). Into the other side of the bay flowed the ancestral Rhine and Meuse rivers, while to the south, where the Channel now is, was dry land, the chalk ridge across the Dover Strait mentioned earlier still being intact.

The deposits at Pakefield clearly relate to an interglacial earlier than the great Anglian Glaciation of around 450,000 BP, which destroyed the Bytham River, but which one? Geological study of the exposed cliff section above the tool-bearing deposits suggests at least two full glacial-interglacial cycles after the Pakefield Interglacial and before the Anglian Glaciation. This would suggest that the Pakefield deposits date at least as early as 700,000 BP (see chart, Fig. 11, p.14). This is confirmed by the species of voles present, some of which went extinct around 600,000 years ago, well before the Anglian. It is also supported by a scientific dating method known as amino acid racemisation (see chapter 7).

It is not known which species of early human made the stone tools found at Pakefield. One possibility is *Homo antecessor*, the species defined by Spanish researchers from the fossils found at Gran Dolina near Burgos, which is also associated with a similar simple flake-based stone tool technology (no handaxes). If this species was well adapted to a Mediterranean-type climate and environment, then this might explain a relatively rapid but brief expansion into northern Europe, perhaps along river valleys such as that of the ancestral Rhine, during an interglacial in which, unusually, a Mediterranean-type climate was found as far north as Britain. However, *Homo antecessor* is only one possibility.

At around this time another human species was evolving, *Homo heidelbergensis*.[4] The species is named after a robust, chinless lower jaw that was found in a sand quarry at Mauer, near Heidelberg, Germany in 1907. (It is quite common for species to be named after the place where they were first recognised but of course this provides no clue to their actual place of origin.) Human fossils attributed to *Homo heidelbergensis* are found in both Europe (including Britain) and Africa. It is generally believed that the species evolved from African *Homo erectus* at least 600,000 years ago (see Fig. 2″3). Although similar to *erectus* in many respects, sufficient distinctive features can be recognised to distinguish *heidelbergensis* as a separate species. Important amongst these are the shape and size of the braincase, which is higher and more filled out than *erectus*, and more closely approaches the average brain size of living humans. Like *Homo erectus*, *Homo heidelbergensis* is often associated with the use of handaxes, and in Britain produced some of the most accomplished and beautiful examples of these tools to be found anywhere (Fig. 26).

After the episode of human occupation around 700,000 BP recorded at Pakefield, there is little unequivocal evidence of human occupation in Britain, or indeed in the whole of northern Europe, until around 500,000 BP (although this is a situation that may well change as a result of on-going research). Around 500,000 BP and afterwards the evidence becomes rich and extensive. Good evidence has now been found from several sites in Britain of occupation that can be dated to the interglacial that im-

---

[4] Some scholars do not accept the distinction between *Homo antecessor* and *Homo heidelbergensis*

Fig. 26: A flint handaxe from Boxgrove, West Sussex, around 500,000 years old

mediately preceded the Anglian Glaciation (MIS 12). This interglacial (MIS 13) belongs to what is known in Britain as the Cromerian, after the town of Cromer on the Norfolk coast.[5]

The most important of these British sites is Boxgrove in West Sussex, near Chichester. Here commercial quarrying for ancient marine sands had for many years turned up handaxes, and from small beginnings a large-scale programme of archaeological excavations developed in the 1980s and 90s. This uncovered one of the most perfectly preserved ancient landscapes known in the world.

Although Boxgrove is now 10 kilometres from the coast, the marine sands were laid down during interglacial conditions when the sea reached further inland than today (Fig. 30). The sea cut a cliff 30 kilometres long and 100 metres high into the chalk hills of the South Downs. However, as the sea gradually retreated, sand gave way to silts and a huge lagoon formed, surrounded by salt marshes and grasslands. Herds of grazers and browsers, such as horse, bison, deer, elephant and rhinoceros, were attracted to the riches this new coastal plain offered. These in turn attracted predators such as wolf, hyaena, lion and *Homo heidelbergensis*. For *heidelbergensis*, this landscape held an added attraction in that the chalk cliffs were an excellent source of high-quality flint from which they could fashion their handaxes, over 300 of which were found during the excavations (Fig. 26).

The actual physical remains of *Homo heidelbergensis* from Boxgrove are not particularly spectacular, just the shaft of a shin bone and two teeth. Even so, such scrappy remains can yield considerable information (and ancient human fossils are so rare that any are very valuable). Study of the shin bone (tibia), which is enormously thick, shows that it belonged to a very heavy and muscular individual, probably a man, perhaps about 5ft 11in in height, who had led a tough, physically demanding life. Microscopic analysis of the two teeth (lower incisors) shows that they bear scratch marks indicating that the jaw was used as a kind of vice to clamp something – hide or meat perhaps – while it was cut with a sharp flint tool.

Nevertheless, it is not for the human fossil remains that Boxgrove is most important but for the archaeology. As we saw in the previous section, most Pleistocene archaeological remains have been subject to a range of natural forces that have caused the remains to be greatly disturbed. Things are not found where they had originally dropped. Such remains are described as having been found in a *secondary context*. Only very occasionally are things found where they originally dropped. Such remains are de-

---

[5] Strictly speaking 'Cromerian' refers to interglacial deposits represented at West Runton in Norfolk (possibly MIS 15) but the term is sometimes used more loosely to describe a whole sequence of marine isotope stages, encompassing four interglacials and three glacials, before the Anglian, going back to the time of Pakefield and beyond. The term 'Cromerian Complex' refers more specifically to this wider usage (see Fig. 11).

Fig. 27: This scatter of flint debris excavated at Boxgrove shows where a hominin sat down to make a handaxe around 500,000 years ago. The sharp boundaries to the scatter show where the knapper's legs were.

scribed as being found in a *primary context*. Boxgrove is a rare example of a primary context site. This arises from the way that the deposits were formed – gently accumulating silts – and were subsequently covered over and preserved.

At Boxgrove scatters of flint flakes have been preserved, almost exactly where they fell, that are the result of a hominin sitting down and making a handaxe (Fig. 27). The flakes can be reassembled into the flint nodule from which the handaxe was made. Only a hole in the middle remains, representing the handaxe itself, which was taken away to be used (see chapter 6, Fig. 128, p.94). Even the tools used to make the handaxes are preserved. These include a red deer antler with flint fragments embedded in it, which had evidently been repeatedly used as a soft hammer to remove delicate flakes.

The evidence from Boxgrove also allows individual butchery events to be reconstructed, as people gathered around the fallen carcass of a horse or rhinoceros and systematically dismembered it (Fig. 28). The cut-marks from their handaxes can be seen on the remaining bones of the butchered animal. In all cases studied at Boxgrove the cut marks from stone tools *underlie* the gnaw marks of carnivores who also had some share of the carcass. This is highly significant because it means that *Homo heidelbergensis* had primary access to the carcass. Others, such as hyaenas, came along and scavenged what was left behind.

This conclusion is important because it suggests that *Homo heidelbergensis* were hunters rather than scavengers. There had long been debate in the archaeological literature about the hunting abilities of archaic humans like *heidelbergensis*. Humans are very poorly naturally endowed as hunters compared with carnivores such as lions. Many believed that they would have been unable to bring down large prey such as rhinoceros or horse with just simple stone tools, and that they lacked the mental capacity for the long-term planning required for complex hunting strategies. It was argued that the 'niche' of early humans was as scavengers rather than hunters, using their stone tools to get access to the parts of carcasses the real predators left behind. The evidence from Boxgrove suggests otherwise.

In order to hunt large prey spears would have been required. As these will have been made of wood, in normal soil conditions (such as occur at Boxgrove) they will decay rapidly and not survive. The only evidence for the use of spears at Boxgrove is an impact fracture in the shoulder blade of a horse that has the characteristics of having been caused by a javelin. However, at Clacton-on-Sea, Essex, the pointed end of a wooden spear made from yew had been found, back in 1911, in a context not too dissimilar in date to Boxgrove (it belongs to the following interglacial, the Hoxnian, around 400,000 BP) (Fig. 29). Some saw this as a digging stick rather than a spear. The clinching evidence has come since 1995, with the excavation of at least nine exceptionally well-preserved wooden spears, of broadly the same date as the Clacton example, from an open-cast brown-coal mine at Schöningen in Germany. These spears, almost all made of spruce, are associated with extensive evidence for horse butchery.

Due to the extraordinary lack of disturbance of the remains found there, Boxgrove provides very rare

Fig. 28: An artist's reconstruction of a group of *Homo heidelbergensis* butchering a rhinoceros at Boxgrove. In the foreground one member of the group makes a handaxe on the spot, whilst another throws stones to keep scavenging hyaenas at bay. The flint-bearing chalk cliffs created by former high sea levels can be seen in the background.

insights into the behaviour and abilities of hominins living half a million years ago, at the scale of events that would have lasted only minutes or hours. Secondary context sites – the vast majority – cannot provide these kinds of insights but contribute to our understanding of human evolution in other ways. They enable us to understand *patterns* of occupation and of the use of landscapes and resources that can be related to the climatic and environmental records.

In interglacial periods when much of the land would have been densely forested, relatively open coastal plains like that at Boxgrove must have presented favoured environments for early humans, and routeways for the colonisation of new lands. River valleys presented similarly favoured environments and routeways. Here too plentiful game would be found, attracted by the river, and large browsers such as elephants, which were probably numerous, would have helped to keep the river banks relatively clear of dense vegetation.

It appears that the Bytham River played a very important role in this respect, and was one of the major routes of human colonisation into Britain before it was obliterated by the Anglian Glaciation of around 450,000 BP. We have already encountered the Bytham River in our discussion of Pakefield, which appears to have been situated in an estuary of this river system, and the very early colonisation of Britain

Fig. 29: A wooden spear tip from Clacton-on-Sea, Essex. It is probably around 400,000 years old.

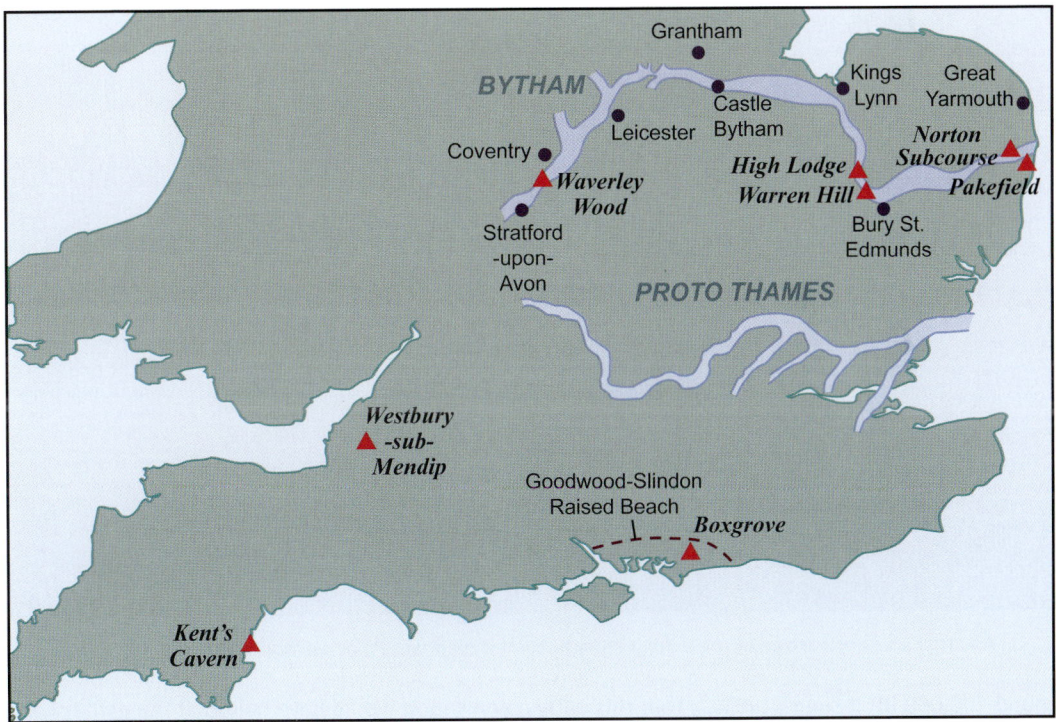

Fig. 30: The reconstructed course of the 'lost' Bytham River. The river was obliterated by the Anglian glaciation around 450,000 years ago. Waverley Wood and some of the other quarries on the course of the river that have produced stone tools and other important Ice Age evidence are shown. The river was 'discovered' (i.e. the evidence was recognised for what it was) in a quarry near the village of Castle Bytham in the 1980s. Also shown are the very early site of Pakefield, which probably lies within the estuary of the Bytham River, and other important pre-Anglian sites elsewhere in the country, such as the quarries at Boxgrove and Westbury-sub-Mendip.

Fig. 31: An artist's reconstruction of the extinct straight-tusked elephant *Palaeoloxodon antiquus*

around 700,000 BP. It seems unlikely that this early colonisation would have halted at the estuary and that occupation did not also occur further upstream. This certainly happened in later interglacials, and particularly perhaps during MIS 13, the last of these interglacials before the Anglian, and the period of the occupation at Boxgrove. By this time some, but perhaps not all, of the colonists were handaxe-using *Homo heidelbergensis*.

A series of quarries are situated along the course of the Bytham, working the sands and gravels of the former river, from Norton Subcourse near to the estuary to Waverley Wood in the vicinity of Coventry, deep in the Midlands and not far from the headwaters of the river (Fig. 30). Through good collaboration between the quarry operators and researchers, many of these quarries have produced (and, at the time of writing, are producing) a wealth of important scientific evidence. Most, but not all, of this evidence comes from secondary contexts, but from this we can nevertheless infer patterns of occupation and early human behaviour through time.

Waverley Wood is of interest for two main reasons. First, it shows colonisation reaching deep inland, and it cannot be an isolated instance. Second, it shows how the colonists, presumably *Homo heidelbergensis*, adapted their technology to changing resources.

Quarrying at Waverley Wood has encountered ancient channels of the Bytham, cut into the underlying bedrock, on two principal occasions, first in the late 1980s and most recently between 2004 and 2006. The sediments from the ancient channels range from the sand and fine gravel of active channels to the organic muds of meander cut-offs. They have produced a range of fossils of vertebrates, invertebrates and plants, enabling detailed reconstruction of the interglacial environment. The most common mammalian fossils are remains of straight-tusked elephants (Fig. 31).

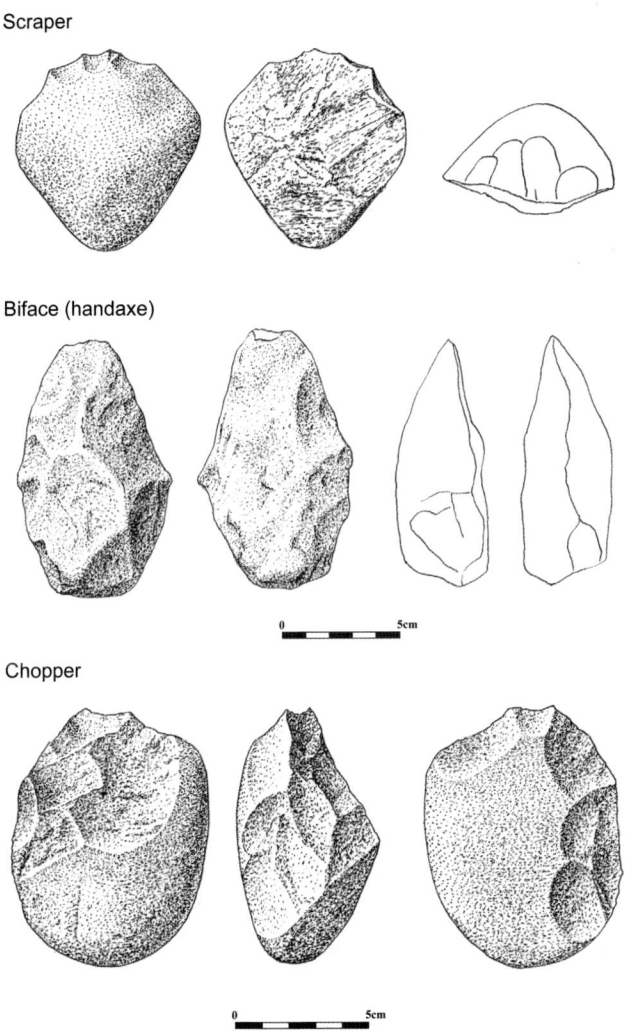

Whenever the quarrying exposes these deposits, stone tools turn up from the floor of the pit. Many of these have been found by the quarry workers, although they are only very rarely found *in situ*. The majority of the tools have been recovered by researchers systematically going through the quarrying 'reject piles'. To date, some seventy tools have been recovered, with a further dozen or so possible items.

Much of the interest of these tools lies in the fact that, with one exception, they are not made of flint. As the ancient colonists moved inland from East Anglia they passed from an area where flint suitable for tool making was readily available to an area where it was not. They had to adapt and make do with other materials, principally locally available quartzite pebbles (Fig. 32). Quartzite can be used to make stone tools but it is much less satisfactory and more difficult to work than flint. The resulting tools are generally rather crude and difficult to recognise to the untrained eye. The majority of the tools recovered are simple choppers and cores for the production of flakes. However, they did attempt to

Fig. 32: A quartzite scraper, handaxe and chopper from Waverley Wood, Warwickshire, around 500,000 or more years old. The crudeness of the handaxe is a consequence of the intractability of the stone rather than the skill of the knapper or the age of the tool.

Fig. 33: Examples of highly accomplished, symmetrical handaxes from Waverley Wood, made of andesite, a fine-grained volcanic rock

make handaxes out of quartzite, although the results were necessarily rather unsatisfactory and crude, and some examples were left unfinished. One can almost follow the thought processes – and share the frustration – of the tool makers as they grappled with the intractable material.

The possession of good-quality, symmetrical handaxes, similar to those found in such abundance at Boxgrove, was clearly important to these people, however, and they sought out an appropriate raw material. What they used was a fine-grained volcanic rock called andesite. The ultimate source of this rock is the Lake District although it is assumed that the handaxe makers obtained it more locally, presumably from blocks transported by an earlier glaciation. Using the andesite the hominins at Waverley Wood made handaxes every bit as accomplished, symmetrical and beautiful as the Boxgrove examples (Fig. 33). Intriguingly, a single, broken, flint handaxe was also found at Waverley Wood. It seems most likely that this was transported by people upstream along the Bytham River as a treasured possession.

Waverley Wood challenges some of the assumptions that have been made about hominins at this stage of human evolution. It is sometimes argued that these hominins lacked forward planning and had what has been described as a 'fifteen-minute culture'. They supposedly made their stone tools from locally available material only as they needed them and then discarded them after use. The evidence from Waverley Wood, where these early humans were near the limit of their range in the Old World, suggests something rather more complex and sophisticated. The study of human evolution is not just, or even mainly, about finding human fossils.

It would be inappropriate here to attempt to trace the patterns of human occupation and activity in Britain through all the interglacial-glacial cycles following the great Anglian Glaciation, after which a new world and new landscapes opened up for colonisation. It would appear, however, that over the long term human evolution in Africa and in Europe took different paths. Slowly a species emerged in Europe that was well adapted, culturally as well as perhaps physically, to the cool climates that were the predominant regime of Ice Age Europe. These were the Neanderthals, *Homo Neanderthalensis*. This species is named after the Neander valley ('thal' in German) where the first Neanderthal bones were recognised, during limestone quarrying, in 1856.

Neanderthals were short and stocky, which may be an adaptation to cold climates as this body shape reduces body surface area and heat loss (Fig. 34). It is found today, for example, amongst the Inuit ('Eskimos'). Neanderthal skeletons display evidence of a tough, physically demanding lifestyle, with evidence of healed wounds and fractures common. The braincase was as large or larger than that of modern humans but very differently shaped, being long and low lacking a rounded forehead (Fig. 35). However, size is not everything, and the capabilities of Neanderthals need to be determined by other means, primarily their behaviour as revealed through archaeology. Above the eye-sockets was a

HUMAN EVOLUTION AND ICE AGE BRITAIN

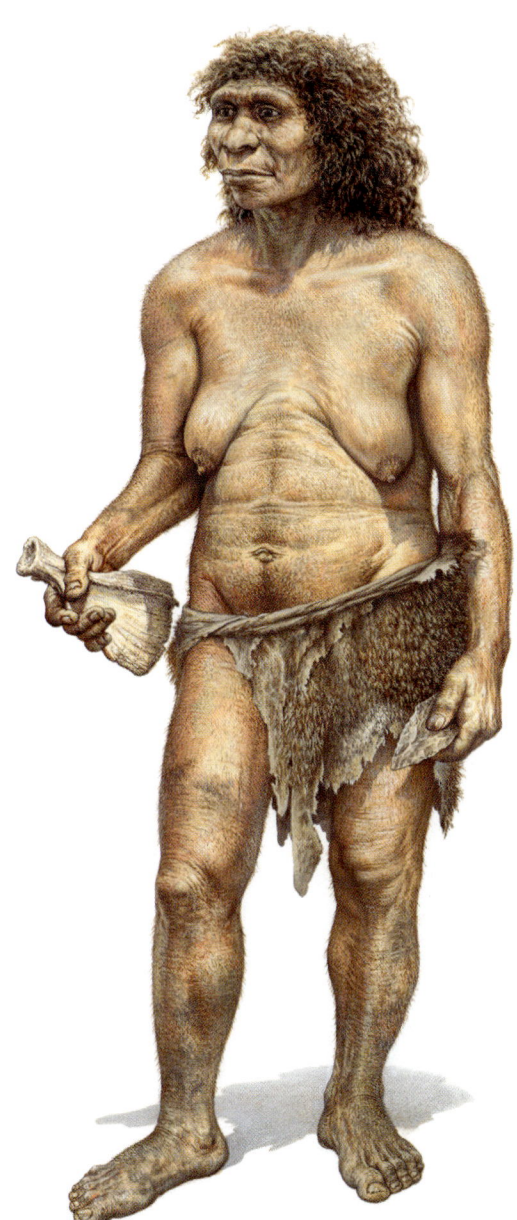

Fig. 34: An artist's reconstruction of a Neanderthal female

pronounced double-arched brow ridge. The face was dominated by a very large nose, perhaps also a cold-climate adaptation, allowing inhaled air to be warmed up. The prominent nose was accentuated by swept-back cheekbones and a receding chin. Although some of the physical features of Neanderthals may be cold-climate adaptations, the archaeological evidence shows that they also occupied warm environments.

The 'classic' Neanderthals occupy a time range from around 70,000 to 29,000 BP but recent research suggests that the evolution of the Neanderthals may be traced back to around 400,000 years ago. A key site for this argument is the Sima de los Huesos ('Pit of Bones'), a remote chamber in a cave system in the Atapuerca Hills of Spain, close to the Gran Dolina which produced human remains dating to some 400,000 years earlier. The pit takes its name from the huge collection of human bones that accumulated in it – it is not properly understood how. Over 4,000 human bones have now been recovered, representing about thirty men, women and children (Fig. 36).

The sizes of the braincases of the Sima individuals fall into the modern range. The long, low skulls with jutting brow ridges over the eye-sockets, share many features with the later Neanderthals. The faces are variable, but some show the pronounced nose so typical of the Neanderthals.

A partial skull from Barnfield Pit, Swanscombe, Kent, a quarry which worked the deposits of the Thames Valley, is of about the same date as the Sima fossils. The three surviving pieces of the skull were found separately, in 1935, 1936 and 1955. Unfortunately only the back and sides of the skull are represented, and the face is missing (Fig. 37). Nevertheless, the partial skull has affinities with both the Sima skulls and later Neanderthal skulls, including the tell-tale peculiarity of possessing a small pit, called the suprainiac fossa, at the back of the skull.

While Neanderthals were evolving, a new type of stone-tool technology appeared around 300,000 BP. It is known as the Levallois technique (Fig. 38), and basically involves striking off a flake of predetermined shape from a carefully prepared core (Levallois technology is described in more detail in chapter 6). The technique, also known as 'prepared core technology', allows much more control over tool production than earlier techniques and was an important innovation. Its introduction marks the

Fig. 35: Neanderthal (right) and modern human skulls (left) compared. The modern human example is of Ice Age date (Cro-Magnon).

Fig. 36: An artist's reconstruction of the 'family' at Atapuerca, northern Spain, around 400,000 years ago, based on the numerous human fossils found in the Sima de los Huesos ('Pit of Bones')

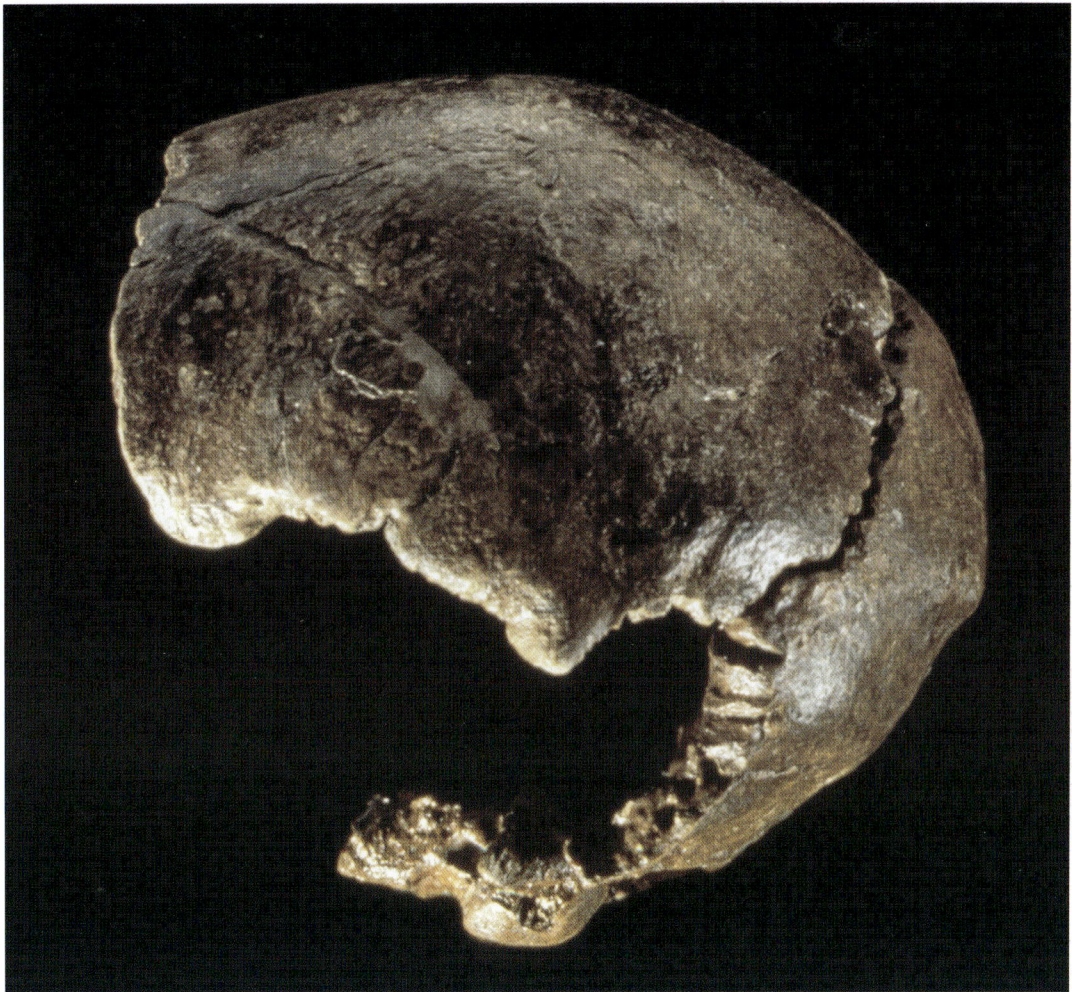

Fig. 37: The partial human skull from Barnfield Pit, Swanscombe, Kent. The skull, which unfortunately lacks the face and jaw, dates to around 400,000 years ago and may represent a transitional stage from late *Homo heidelbergensis* to early Neanderthal.

beginning of the Middle Palaeolithic, the second major division of the Old Stone Age, succeeding the Lower Palaeolithic.

In Europe the Levallois technique is associated with the Neanderthals, whose varied stone-tool industries are called Mousterian after the French rockshelter site of Le Moustier. Mousterian tool kits frequently contain a high proportion of tools made using the Levallois technique. However, the Levallois technique is not unique to Neanderthals, and in western Asia and Africa the technique was also used by early populations of *Homo sapiens*.

In Britain, the only Neanderthal fossil remains known are seventeen teeth from Pontnewydd Cave in North Wales, dating to around 225,000 BP (MIS 7, the Aveley Interglacial). Their stone tools, however, are found in some numbers. At Pontnewydd, for example, hundreds of stone tools were found in the deposits containing the early Neanderthal teeth, including small pointed handaxes and Levallois cores and flakes. Some of the tools were burnt, suggesting they had lain in a hearth, indicating the controlled use of fire.

Fig. 38: Levallois and other Middle Palaeolithic flakes, made from a range of different stone types

The advent of the ability to make or control fire is of course a critical event in human evolution. Fire provides warmth, protection against wild animals and, through cooking, the ability to make indigestible foods digestible and protect against disease. It has been described as providing humans with a kind of external second stomach. However, in the absence of formally constructed hearths, which only make a late appearance in the record, it is difficult to date the appearance of the controlled use of fire. This is because of the difficulty of distinguishing natural, wild fires from controlled, 'domesticated' fires. Despite this, there is now convincing evidence from a number of sites in Europe, most notably Beeches Pit in Suffolk, for the controlled use of fire from around 400,000 BP. That is the time of the human fossil remains from La Sima de los Huesos and Swanscombe, as well as the wooden spears from Schöningen (this quarry has also produced convincing evidence of the controlled use of fire). So fire was part of the Neanderthals' cultural tool kit from the outset, opening up possibilities for the occupation of colder environments.

Neanderthal occupation of Britain seems nevertheless to have been intermittent. Several sites along with Pontnewydd Cave have yielded evidence of occupation during the Aveley interglacial (MIS 7) of around 242,000–186,000 BP, but following this interglacial the climate entered into a long and severe glacial (MIS 6), which eventually, current evidence suggests, led to the abandonment of Britain. The declining climate did not, however, drive the early Neanderthals away immediately, an indication of their abilities for cold-climate adaptation. At Crayford in Kent, for example, Levallois stone tools have been found in association with the jaws of woolly rhinos.

The abandonment of Britain appears to have lasted through the MIS 6 glacial, beginning around

Fig. 39: A map of Britain/'Doggerland' and the 'Channel River' system as it may have looked around 50,000 years ago. Some key sites dating from between about this time and the end of the Pleistocene are shown.

186,000 BP, the succeeding Ipswichian Interglacial (MIS 5e), and the early part of the Devensian cold stage (MIS 5d–4) that followed. On current evidence it was not until towards the beginning of MIS 3, around 60,000 BP, that the Neanderthals returned to Britain. The climate during MIS 3 was predominantly cool rather than very cold or warm, but it was, as the ice-core records show, extremely variable. MIS 3 (c.57,000–24,000 BP) is a very interesting period in terms of human evolution because it witnessed, in its second half, both the extinction of the Neanderthals and the first appearance of *Homo sapiens* in Europe.

Explaining this long period of the apparent abandonment of Britain, some 130,000 years in duration, is difficult. Abandonment during the coldest phases, such as the depths of the MIS 6 and MIS 4 cold stages is unproblematic (although MIS 4 was less severe). The final breaching of the chalk ridge across the Dover Strait occurred during MIS 6 if not earlier. So during the subsequent Ipswichian Interglacial (MIS 5e), when the average temperatures were a few degrees warmer than today and the hippopotamus was common throughout the country, Britain was an island. This may have been sufficient to prevent its reoccupation by Neanderthals. Even as the climate cooled and sea levels fell, the breaching of the chalk ridge led to the formation of a massive river system, the Channel River, fed by such tributaries as the Thames and the Rhine, which would have formed a formidable barrier (Fig. 39).

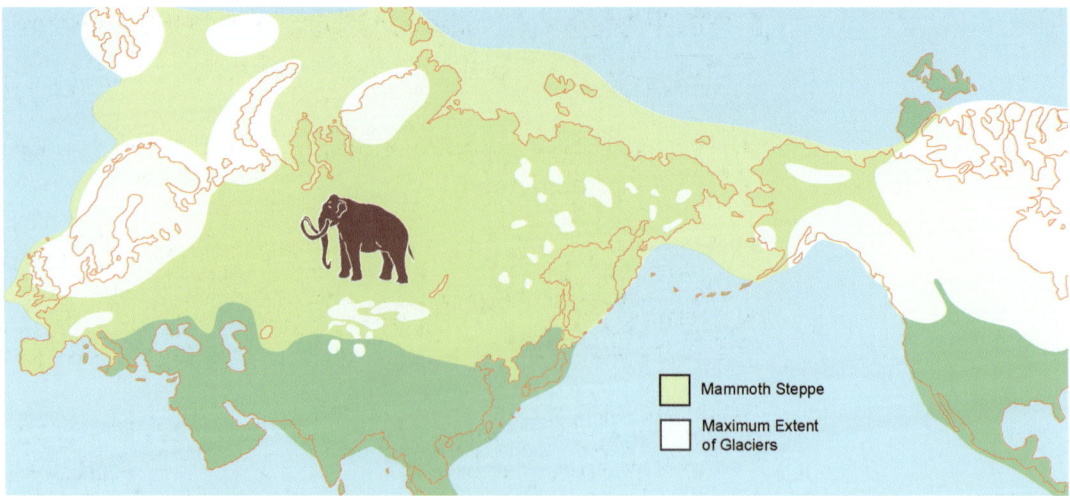

Fig. 40: A map of the Mammoth Steppe, which stretched uninterrupted from Britain to Alaska and included the Ice Age 'lost lands' of *Doggerland* (between Britain and continental Europe) and *Beringia* (between Siberia and Alaska)

Fig. 41: A flint handaxe in the course of excavation at Lynford Quarry, Norfolk. This small axe is typical of the type used by Neanderthals.

When the Neanderthals returned to Britain around 60,000 BP, it was into the environment of the *Mammoth Steppe*, which would have been familiar to them. The Mammoth Steppe was an extraordinary phenomenon, with no real parallel today. It stretched, uninterrupted, half way around the top of the world, from Britain in the west, across Doggerland (the landscape now flooded by the southern North Sea), the North European Plain and Siberia, all the way to Alaska in the 'east' (Fig. 40).

A very important Neanderthal site on the Mammoth Steppe was uncovered in 2002 during aggregate extraction at Lynford Quarry in Norfolk. The quarrying uncovered an ancient river meander channel, filled with fossil bones and flint artefacts. The site dates to the beginning of the reoccupation of Britain, around 60,000 BP, a brief period of relative warmth during the very variable climate of MIS 3.

Archaeological excavation of the organic deposits in the river channel produced a wide range of superbly preserved fossils, including the remains of plants, beetles, molluscs and vertebrates. Together these allowed the local environment and climate to be reconstructed. It was an open environment, dominated by grassland with small stands of arctic dwarf trees. Summer temperatures would have been in the range 12–14°C, with winter temperatures at or below minus 10°C.

The excavations produced over 2,700 flint artefacts, including forty-seven beautiful examples of the characteristic small handaxes used by Neanderthals at this period (Fig. 41). Large animal remains from the deposits included reindeer, woolly rhino, horse, bison, brown bear and wolf. Woolly mammoth was best represented, however, with the remains of at least eleven mammoths, mostly adults, being identified (Fig. 42).

Fig. 42: An adult mammoth tusk under excavation at Lynford Quarry, Norfolk

The association of the flint tools with the animal remains, especially the mammoths, is unlikely to be fortuitous. One interpretation is that the Neanderthals selected vulnerable individuals and drove them into the river channel where they would become mired and could be dispatched at close range using spears (Fig. 43). This interpretation is supported by the fact that detailed study of the mammoth body parts represented shows that the meatiest large limb bones are missing. Presumably they were taken elsewhere to be eaten, with the remainder of the carcass left behind. That this was done by Neanderthals rather than by carnivores such as the hyaena, is indicated by the near absence of gnaw marks.

Lynford opens up a window into Neanderthal lifestyle and abilities. It suggests well organised, thoughtful, collaborative hunting. For groups to survive in such bleak environments, relatively developed social and cultural adaptations would have been necessary. The latter must have included warm clothing made of animal skins, although direct evidence for this has not survived. However in Africa at this time, and already beginning to spread beyond it, was a species with even more advanced abilities – *Homo sapiens*, modern humans.

Fig. 43: Artist's reconstruction of Neanderthals butchering a mammoth mired in the river at Lynford, Norfolk

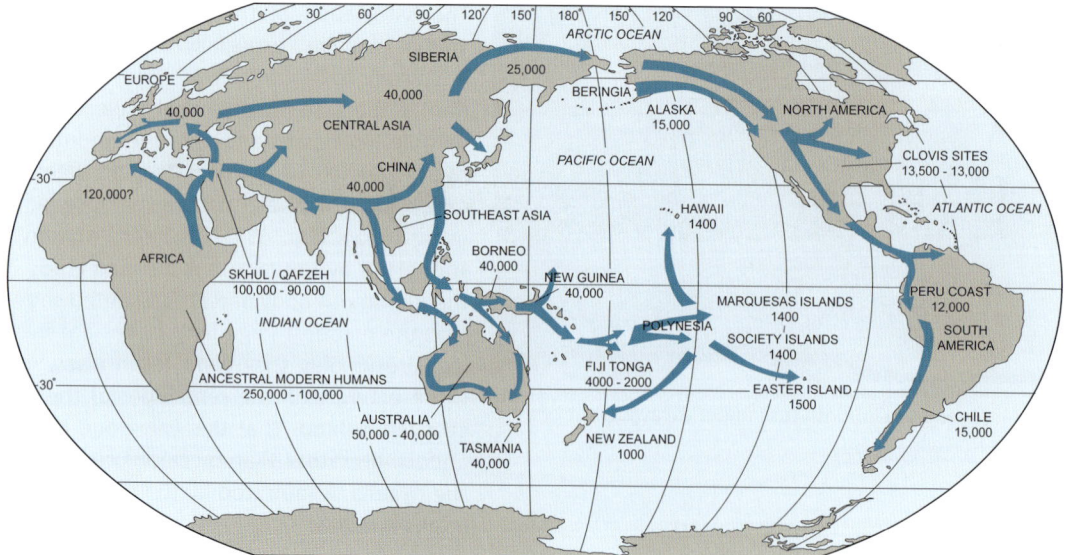

Fig. 44: The colonisation of the world by modern humans (*Homo sapiens*). Approximate dates in years BP.

While in Europe we appear to witness from around 400,000 BP a long process of 'Neanderthalisation', so in Africa from around the same time we appear to witness a long process of 'modernisation'. Both Neanderthals and modern humans probably shared a common origin in *Homo heidelbergensis* but in the different continents and environments of Europe and Africa their evolutionary paths diverged (Fig. 23, p.27). In both cases we see changes in both the skeleton, notably the skull, and in behaviour, as revealed by the archaeological record. The braincase enlarged in both to present-day values but the shape of the modern human skull is very different to that of the Neanderthal. The modern human skull is loaf-shaped with a high forehead, browridges are reduced, a small face with a small nose is 'tucked in' below the forehead, and the lower jaw develops a distinct chin. Characteristics such as these begin to appear in the (patchy) African fossil record from before 150,000 BP.

The most important differences are, however, in behaviour. With *Homo sapiens* we begin to see evidence for the first unequivocal, deliberate production of art objects. Stone tool types are more varied, and elaborate tools are made from bone and antler, something the Neanderthals did not do. Other changes, such as new and more flexible ways of making a living, exploiting a wider range of wild resources (including those of the seashore), can also be teased from the archaeological record.

Amongst the clearest indication of the behavioural flexibility of *Homo sapiens* is their spread into a wide range of different environments including, from perhaps around 60,000 years ago, into southern Asia and on to Australia (Fig. 44). This was the first time Australia had been colonised by humans, and even at times of low sea level the crossing would have involved a sea journey, implying the use of a craft of some kind.

Modern humans first colonised Europe around 45,000 years ago, where they would have encountered the Neanderthals. In Europe, this colonisation is associated with dramatic changes in the archaeological record, including the appearance of portable art (figurines etc.) and cave art, sophisticated bone and stone tools and elaborate burials. These developments (not all of which occurred at precisely the same time) mark the beginning of the Upper Palaeolithic, the last major division of the Old Stone Age.

Such developments also mark the beginnings of religious belief in any sense that we would understand it – perhaps the most ubiquitous and characteristic attribute of *Homo sapiens* that can be found. In

Fig. 45: A statuette sculpted in mammoth ivory, from Hohlenstein-Stadel, Germany. The figure is half human, half lion and dates to around 34,000 BP.

Fig. 46: Artist's reconstruction of the burial of the 'Red Lady', Paviland Cave, Gower, Wales, about 32,000 BP

south Germany, for example, a figurine made of mammoth ivory has been recovered from the Hohlenstein-Stadel cave, dating to around 34,000 BP (Fig. 45). It depicts a creature with a human body and a lion's head. Is it a god, a mythological creature or a shaman? While we cannot answer this question, the figurine is undoubtedly evidence for spirituality.

Until quite recently it was fairly widely held that all these developments originated fairly rapidly around 50,000 BP in an 'explosion' of culture that marked the advent of fully modern behaviour. This was explained by a speciation event, triggered by a genetic mutation (perhaps a mutation that enabled the development of complex grammatical language for the first time), which brought behaviourally modern *Homo sapiens* rapidly on to the scene. However, work in Africa, particularly over the past decade, is increasingly undermining this view. Signs of modern human behaviour, including the production of sophisticated stone tools and the use of pigments (for body decoration?), appear sporadically in the African record from before 200,000 BP. From after 100,000 BP the evidence becomes stronger and more varied, including the appearance of barbed bone points, evidently harpoons, and beads for personal ornament. A critical site is Blombos Cave in South Africa, where shell beads and pieces of red ochre engraved with regular criss-cross patterns – the earliest known art – are dated to around 74,000 BP. This evidence suggests that the process of becoming behaviourally modern was complex and drawn out.

The nature of the interaction between *Homo sapiens* and *Homo neanderthalensis* in Europe, and the cause of the Neanderthals' extinction, have been topics of perennial fascination for scientists and novelists alike. The Neanderthals became extinct between about 35,000 BP and 29,000 BP and it seems too much of a coincidence to most people that this is the same period when modern humans were becoming firmly established in Europe. However, that does not mean that there was warfare between these two human populations in which the Neanderthals were eventually wiped out, although this possibility cannot be ruled out. As both species were at root very similar, and were trying to occupy the same 'niche' as hunters, *Homo sapiens* with their more flexible, creative behaviour may have simply out competed the Neanderthals, who were driven slowly to extinction. Another factor may have been the very unstable climate around 30,000 BP, which challenged Neanderthal groups beyond their limit but to which the more flexible, socially more complex and culturally more sophisticated modern humans were able to adapt successfully.

Homo sapiens first colonised Britain perhaps around 44,000 BP. The Upper Palaeolithic in Britain is divided in two by the last major advance of the ice sheets, the peak of the Devensian glaciation (MIS 2). During the period of this ice advance, the Last Glacial Maximum (LGM) between about 27,000 BP and 16,000 BP, Britain was once again abandoned by humans. It was too bleak and barren even for adaptable *Homo sapiens.* The period before the abandonment is known as the *Early Upper Palaeolithic* (c.44,000–30,000 BP) and the period after it is the *Late Upper Palaeolithic* (c.15,000–14,000 BP).

Human fossil remains from the Early Upper Palaeolithic are very rare, and occupation during this period is mainly recognised from characteristic stone tool types, which now show much more regional and chronological variation than in earlier periods (another characteristic of modern human behaviour). However, a very important site from this period is Goat's Hole Cave, Paviland, in South Wales. Around 32,000 BP an elaborate burial was made in the cave, one of the earliest known from Europe (Fig. 46). The body was covered with red ochre and accompanied by numerous grave goods, including mammoth ivory bracelets and rods, and small beads made from perforated periwinkle shells. The burial was excavated in the 1820s and was dubbed the 'Red Lady of Paviland' (with deliberate overtones of prostitution) on account of the ochre staining. Modern analysis of the bones has shown the burial to have been that of an adult male.

Following abandonment during the Last Glacial Maximum, Britain was recolonised again by modern humans around 15,000 BP. Although climate was ameliorating, sea levels were still low and the vast expanse of Doggerland connected modern-day England with the Low Countries. Doggerland was more than just a 'land bridge' between continental Europe and Britain, it was a rich landscape of river valleys, plains, coasts and marshes, teeming with wildlife. Many people must have lived and died in these now drowned landscapes.

During the Late Upper Palaeolithic the main evidence of occupation again takes the form of characteristic types of stone tools. Over much of continental Europe a very widespread cultural tradition known as the Magdalenian is found from around 22,000 BP to around 14,000 BP. It is named after La Madeleine, a rock-shelter in France. In Britain, from 15,000 BP the principal cultural tradition is called the Creswellian, after the caves at Creswell Crags, a limestone gorge on the Derbyshire/Nottinghamshire border where many tools of this tradition have been found. The Creswellian appears to be a late regional variant of the Magdalenian. Tools such as scrapers and Cheddar Points (named after the caves of the Cheddar Gorge) were made from long flint blades. Barbed harpoons, needles and batons, all similar to continental examples, were made from antler and bone (Fig. 47).

Most of the famous cave paintings of France and northern Spain, such as those of Lascaux and Altamira, were made during the Magdalenian period. Until recently it was believed that such cave art was absent from Britain. However, in 2003 incised figures were discovered on the walls of Church Hole cave at Creswell Crags (Fig. 48). They are difficult to see except in the right lighting conditions and had been overlooked for more than a century. The engravings include figures of deer, horse, bison and birds that can be linked stylistically with examples from the continent.

Gough's Cave, in the Cheddar Gorge, Somerset, is perhaps the site that best exemplifies the lifestyle of *Homo sapiens* in lateglacial Britain. Excavations have been carried out at the cave since the late 19th century, although unfortunately the earlier excavations were largely 'clearance' operations and much

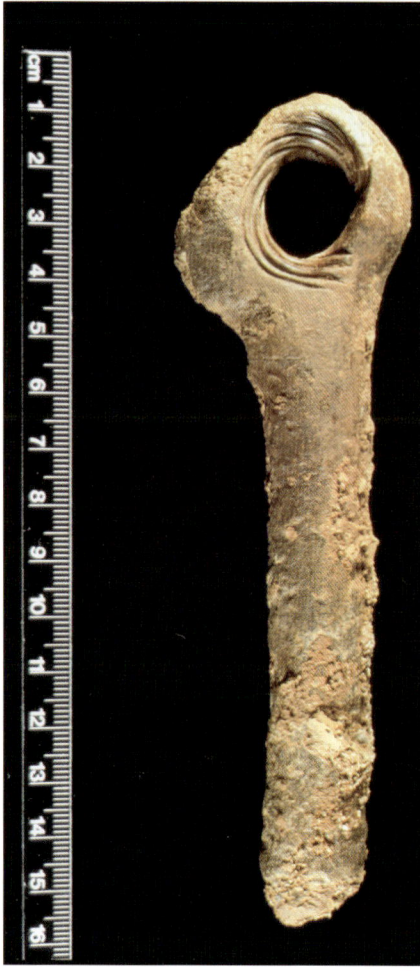

Fig. 47: Late Upper Palaeolithic artefacts. Creswellian tools made on blades, and a reindeer antler baton from Gough's Cave, Cheddar Gorge, Somerset.

archaeological material has been lost. Nevertheless, thousands of animal bones and flint tools of the Creswellian tradition, as well as tools made of ivory, antler and bone, have been recovered. The latter include what appears to be a 'tally-stick' – a piece of animal rib with a series of carefully made scores along its edges (Fig. 49). The animals represented by the bones include red deer, lynx and arctic hare, as well as birds such as black grouse, ptarmigan and whooper swan. By far the largest number of bones, however, are those of horse, and it seems that the hunting of horse was the major activity carried out from the cave.

Around 15,000 BP, when the cave was occupied, the cave would have led out directly into the Gorge, and would have provided an excellent location from which to launch the ambush of wild horses driven into the Gorge (Fig. 50). Perhaps the hunters were assisted by dogs because amongst the animal bones recovered are those of 'wolves' considerably smaller than the Ice Age norm. These are therefore likely to be dogs, which are at root domesticated wolves, selective breeding having led to a reduction in size and other useful qualities such as submissiveness.

The horse bones show extensive evidence of skinning and butchery, such as cut marks from stone tools and breakage to extract marrow. It appears that little of the carcass was not put to good use. Tendons and ligaments were stripped, presumably to make thread and rope, and even the hooves were removed, perhaps to make glue.

A rather gruesome aspect of the animal bones recovered from Gough's Cave is the inclusion of human bones representing at least five individuals: a young child, two adolescents and two adults. Study of the bones suggests that these had been treated in the same way as the animal bones – they carry cut-marks from stone tools and breakage for marrow extraction, for example (Fig. 51). This suggests cannibalism (which may be done for ritual purposes) or ritual dismemberment as part of funerary rituals.

Fig. 48: Bison engraving, Church Hole cave, Creswell Crags, Derbyshire, dating to around 15,000 BP

Fig. 49: The enigmatic 'tally stick' (animal rib) from Gough's Cave, Somerset

Fig. 50: An aerial view of the Cheddar Gorge, Somerset. Caves in the sides of the gorge provided an excellent base for the ambush of wild horses.

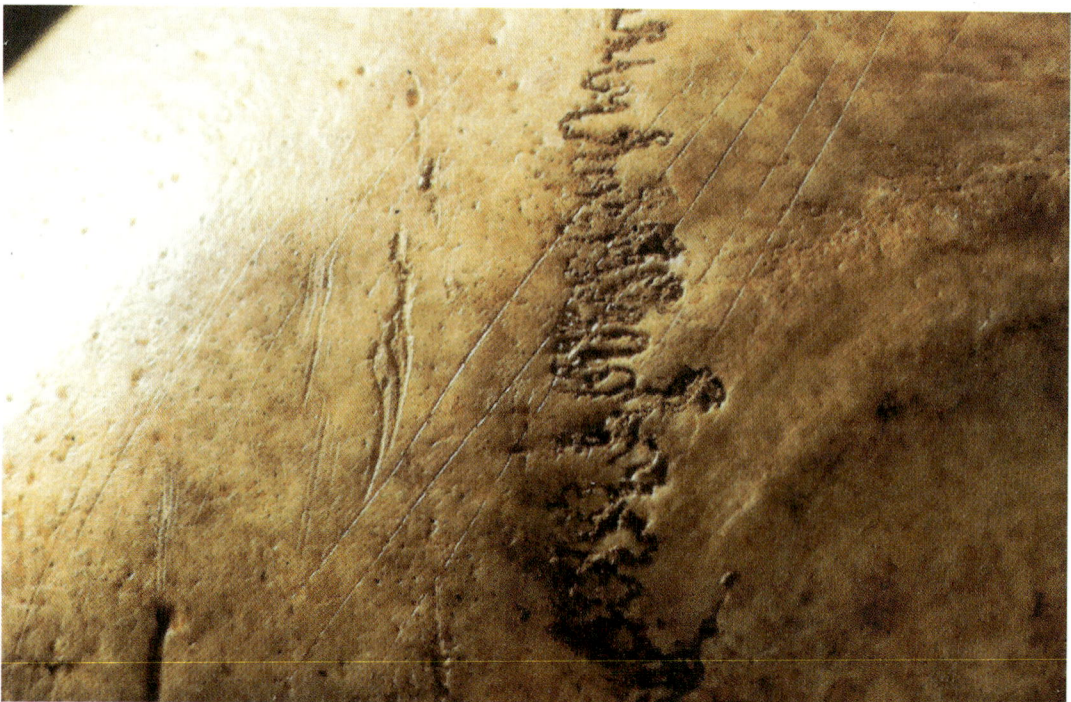

Fig. 51: A human skull from Gough's Cave. The skull bears a series of cut marks made by flint tools, indicating deliberate defleshing.

In the complexity of their lifestyle, the hunters occupying the caves in the Cheddar Gorge or Creswell Crags show a range of behaviours that compares with that of traditional hunter-gatherer societies of the present or recent past. Except for minor details, human evolution was essentially over by the end of the Pleistocene. All subsequent developments, including the technological sophistication of present-day post-industrial societies, are likely to be more a matter of *cultural* rather than *biological* evolution.

Culture and climate continued to change during the final stages of the Ice Age (Fig. 52). In the last part of the relatively warm interstadial following the Last Glacial Maximum, the Creswellian is replaced by new types of stone tools, including distinctive 'penknife points', part of a wider cultural tradition also found in the adjacent lands of continental northwest Europe across the still exposed Doggerland. These new tools are assigned to the Final Upper Palaeolithic, from around 14,000 BP to 12,900 BP. However, the Ice Age was not quite over and there was one last spasm of extreme cold, called the Younger Dryas, lasting between about 12,900 BP and 11,400 BP. During this period the occupation of Britain may have been intermittent, and to this and the beginning of the warm interglacial that followed, belong stone-tool assemblages assigned to the Terminal Upper Palaeolithic, including characteristic 'long blade' types. Around 11,400 BP the climate warmed rapidly, ushering in our present interglacial period, the Holocene. From this point onwards the human occupation of Britain has been unbroken, although it was to take almost a further 5,000 years for Doggerland to fully flood and for Britain to become an island for the last time.

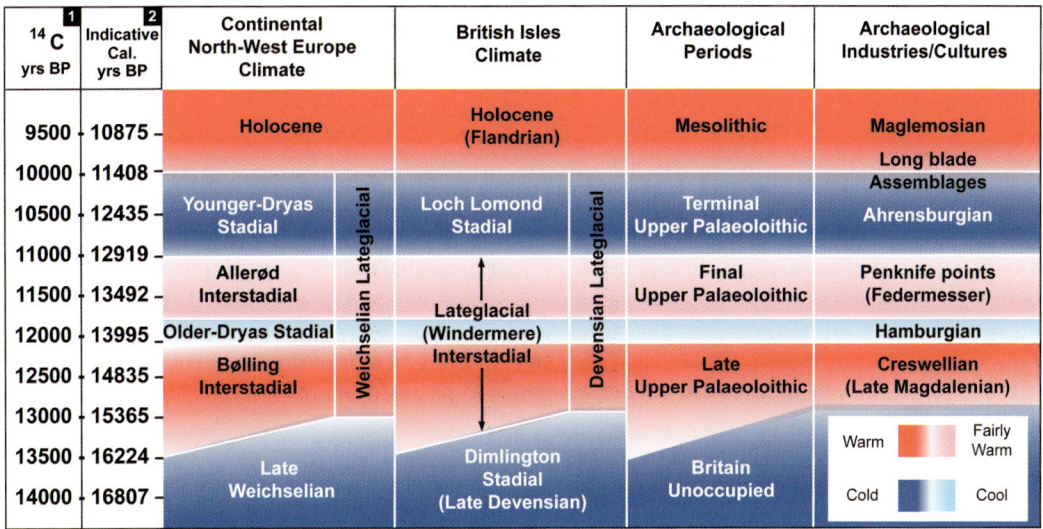

Fig. 52: A chart showing the complex sequence of changing climate and archaeological cultures characterising the final stages of the Pleistocene. The predominance of cultural labels derived from what is now continental Europe reflects the fact that Britain during this period was still a peninsula rather than an island. Note the wide discrepancy between the uncalibrated radiocarbon dates (1), which are used in much of the archaeological literature, and the indicative calibrated dates (2) used in this book: see chapter 7 for an explanation.

Chapter 4: What Remains? – Sediments

As we have seen, much of what we know about the British Quaternary has come about as a result of sand and gravel quarrying exposing Ice Age river sediments and the fossils they contain. In the early days of the quarry industry and well into the 20th century, aggregates were extracted and sorted by hand (Fig. 53). The comparatively slow speed of operation meant that many finds, such as handaxes and ancient bones, were spotted as the gravels were dug out. Today, machines have taken over many of the roles previously done by hand (Fig. 54). Whilst this has increased the productivity of the quarrying industry, it can mean that lots of Ice Age evidence is lost as sediments are rapidly removed.

Unlike much traditional archaeology, recording these Ice Age finds can usually be done quickly during sand and gravel extraction. Ice Age researchers are often interested in documenting sediments to understand Ice Age landscapes. These investigations can be completed rapidly, generally taking between a few hours to one day. Researchers work with the quarry management to minimise disruption as they document exposed sediment sections (Fig. 55). Any finds spotted during the researchers' visit or by quarry workers are noted, their location recorded and, if safe to do so, they are removed to a secure location (see chapter 8 for more detail on these procedures).

Bulk sediment samples can be collected from potentially fossil-rich deposits (described further below) such as fine- grained sediments or organic deposits. These samples tend to be about 20 litres in size and can be used in a variety of palaeoenvironmental investigations through the recovery of small ver-

Fig. 53: Quarrying methods of the early 19th century. A painting by John Linnell of the Kensington Gravel Pits, 1811-12.

Fig. 54: Gravel quarrying in the early 21st century. Whitemoor Haye Gravel Quarry, Staffordshire, 2004.

Fig. 55: Specialists examining an exposed section in a quarry and taking samples

tebrate bones, invertebrate remains and plant fossils. For this evidence to be most useful, the location and depth of these samples needs to be recorded. Once collected, they can then be assessed off-site by appropriate experts.

This chapter provides an outline of the different kinds of sediments that can be found in sand and gravel quarries, how they formed, how they are studied and what they can tell us. The following two chapters deal with the two major classes of finds that can be contained within sediments – fossils and stone tools.

The sediments exposed by quarrying were deposited and formed by three main agents: water, ice and wind. These agents are responsible both for processes of deposition and for erosion of the land surface. A quarry is literally full of evidence for the environments of the Ice Age. Every bucket of sand and gravel removed from a quarry contains sediments that can tell us about the agents which laid them down upon a former land surface.

The basic principles of analysing sediments are very simple and are based upon the size, type, sorting and shape of the sediment components.

The *particle size* of the sediment is one of the most fundamental criteria and allows inferences to be made as to the energy required to move the sediment particles.

The main sediment particle sizes are described in the table below and illustrated in Figure 56. The size of small particles is measured in microns (μm).

Sediment Type	Size	How to identify the sediment
Clay	<2μm	Place a little of the sediment in your palm, spit on the sediment and roll into a sausage. If you can roll the sediment into a sausage and bend it without cracking, then the sediment has a significant clay content.
Silt	2–64μm	Rub a little sediment between your fingers. If there is a silt content, the sediment will feel 'soapy'.
Sand	64μm–2mm	Sand can range from fine to coarse.
Gravel	2–4mm	Small stones, generally half the width of your little fingernail and bigger.
Pebbles	4mm–6.4cm	Stones generally the size of your fingernail and bigger.
Cobbles	6.4cm–25.6cm	Stones the size of your fist and bigger.
Boulders	25.6cm and above	Stones the size of your head and bigger.

The particle size of the sediments determines the force required to move them. For example, a powerful river or a glacier is required to move and re-deposit sediments composed of cobbles and boulders. Conversely, a small stream often has only enough power to move small particles such as sand, silt and clay.

It is also important to characterise the main *rock types* represented in the sediments. This can provide important information about the source of the sediments. For example, quarrying in East Anglia has exposed vast expanses of gravels that contain significant numbers of stones derived from the Midlands and Wales. These were ultimately deposited many miles from where they were originally taken up and

WHAT REMAINS? – SEDIMENTS

Fig. 56: Different sediment types and their sizes, from sand to boulders

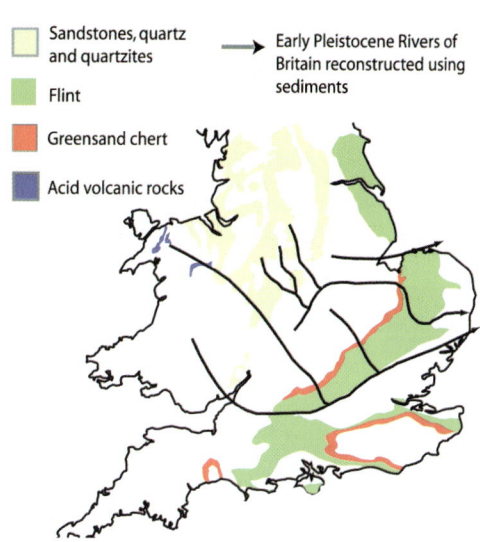

Fig. 57: Map showing reconstructed routes of ancient Pleistocene rivers in relation to geological deposits

transported by a powerful river, the Bytham, which once flowed across the Midlands and out into the sea through East Anglia (Fig. 57).

The following table provides an identification list of common rock/mineral types found in river gravels in Britain. Some of these are illustrated in Figure 58. Figure 59 illustrates rock

Rock type	Colour	Description
Sandstone	Greyish yellow to brown	Gritty, crumbly stone made up of grains of sand.
Quartzite	Typically grey, brown or pink	Hard, metamorphic rock that was originally sandstone. Irregular fractures.
Quartz	Typically semi-transparent, yellow or milky white	Hard crystalline mineral with fractures across a plane.
Flint	Typically black, grey or brown	Smooth texture. The outer 'skin' of flint can become white through weathering over time.
Chert	Varies greatly in colour from white to black but most often grey, brown, greyish brown and light green to rusty red	Fine-grained rock that may contain small marine fossils.
Limestone/ Chalk	Grey rocks	Made up of calcium carbonate that often contains marine microfossils (chalk is white when fresh).
Basalt	Grey to black volcanic rock	Fine-grained due to rapid cooling of lava.
Dolerite	Grey to black volcanic rock. From N. England and Scotland	Similar to basalt but cooled more slowly and so has coarser crystal grains that can be seen with a hand lens.
Rhomb porphyry	Red-purple coloured	Very hard igneous rock consisting of large white crystals set in a finer-grained matrix. From Norway.

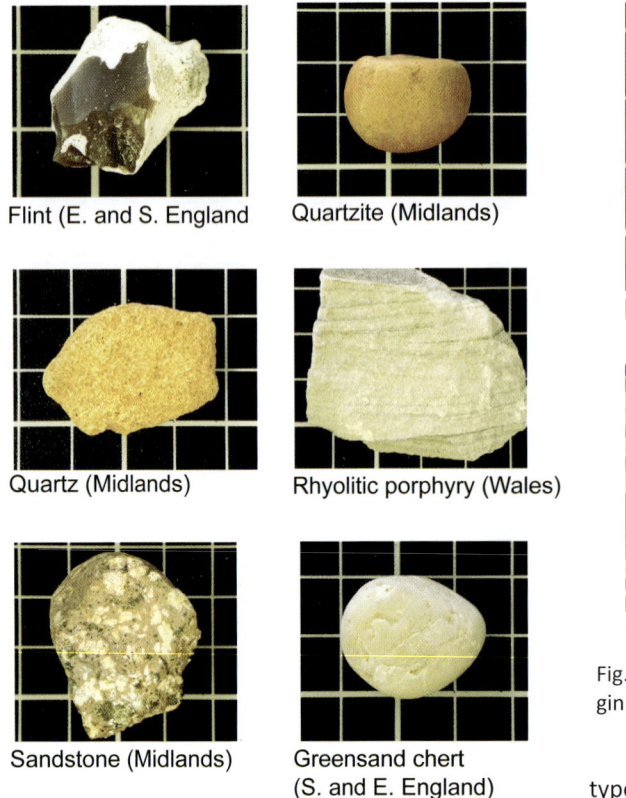

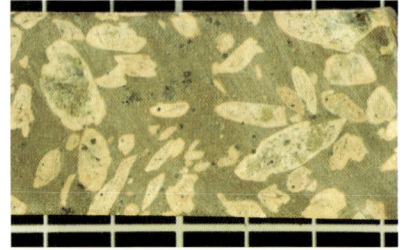

Fig. 59: Some rocks that indicate the former origin of ice flow

Fig. 58: Some of the main rock/mineral types found in British river gravels, and their origin (scale: one grid square width = 1 cm)

types that indicate the place of origin of the ice flow.

The *shape* of a particle, and the extent to which the edges have been rounded, can indicate how it has been affected by processes such as stream flow, exposure to wind or glacial movements. To determine the roundedness of particles objectively, it is best to compare them with a standardised diagram, like the one in Figure 60.

Another factor that is important to consider is the *sorting* of the sediments (Fig. 61). This refers to

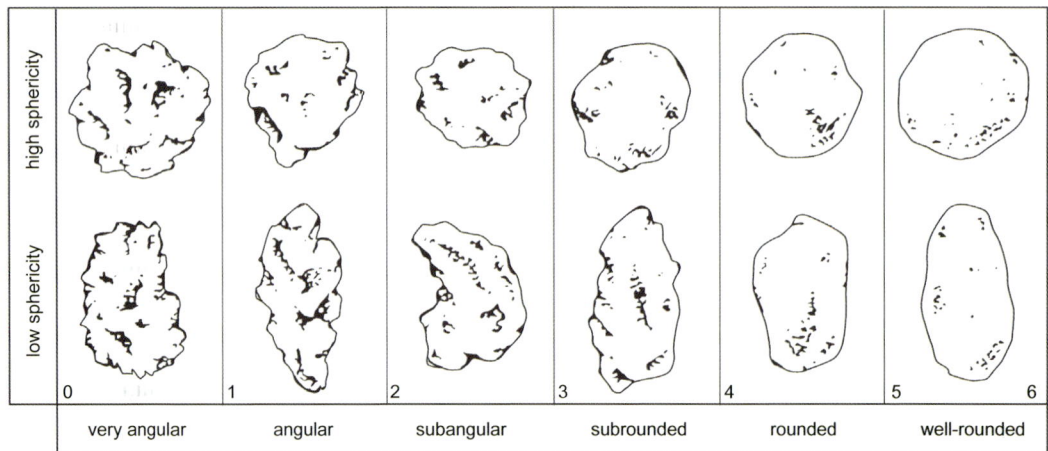

Fig. 60: Diagram for assessing the roundness of sediment particles

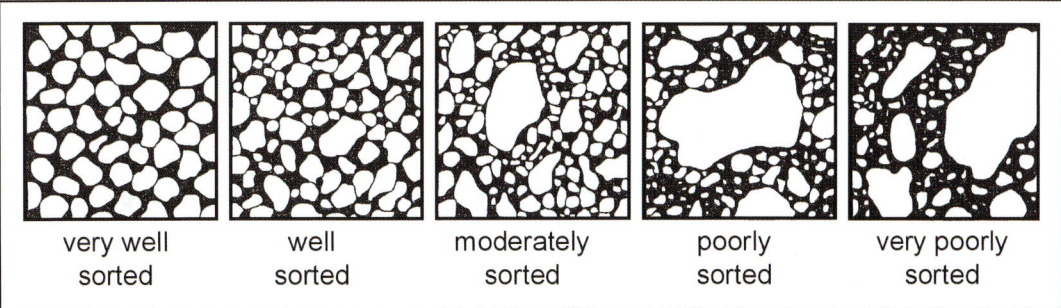

Fig. 61: Diagram for assessing the degree of sorting of sediments

whether they are sorted into layers of predominantly one size or not. This can provide information about the character of the ancient energy source responsible for the deposition of the sediments. Water and wind can be sufficiently variable in strength to sort the sediments according to size, resulting in distinct layers of clays, silts, sands and gravels. A glacier on the other hand picks up everything in its path regardless of size and dumps the resulting mix where it finally melted. Such sediments are very poorly sorted and are described as diamictons.

The main kinds of sediment units that may be encountered in a sand and gravel quarry can be summarised as follows:

GLACIAL TILL

Till or 'boulder clay' is typically a deposit of clay that is full of boulders (i.e. the deposit is poorly sorted) (Fig. 62). Till forms in and beneath ice sheets and glaciers (Fig. 63). As the till is the result of abrasion

Fig. 62: Typical poorly-sorted glacial till with pebbles in a clay matrix

Fig. 63: A modern glacier ploughing through the landscape, picking up and mixing the sediments in its path

of the older rocks over which the ice has travelled, it takes its colour from them. Thus, in Britain, over Triassic and Old Red Sandstone areas boulder clay is red, over Carboniferous rocks it is often black, over Silurian rock it may be beige or grey, and where the ice has passed over chalk the clay may be quite white and chalky (this is known as 'chalky till'). The boulders may be angular, sub-angular or well-rounded, and frequently bear grooves and scratches caused by contact with other rocks while they were held firmly in the moving ice. Like the clay in which they are contained, the boulders originated in the various districts over which the ice has travelled. Boulders transported far from their place of geological origin are known as erratics. By the nature of the contained boulders it is often possible to trace the path along which a vanished ice-sheet moved. For example, in the till of the east coast of England many Scandinavian rocks (erratics) can be recognised.

Stones within a till are often preferentially aligned in the direction of the past of ice-flow. The orientation and dip of the stones are measured to reconstruct ice-flow directions.

Marine Deposits

As described in chapter 2, sea level in Britain has varied dramatically during the Quaternary in response to climate change and the cyclical growth and retreat of ice sheets in northern Europe. Quaternary marine deposits in Britain are typically found in coastal areas and are a testament to higher sea levels in the past. The sea created what are now raised beaches, sometimes several miles inland, and preserved because of subsequent uplift of the land (Fig. 64). Beach deposits can be recognised by very well-sorted and well-rounded pebbles. Beach pebbles are typically covered in 'chatter-marks' due to

Fig. 64: Raised beach sediments at Brighton, south England

Fig. 65: A 'chatter-marked' flint pebble

agitation caused by tidal action (Fig. 65). Marine fossil shells preserved in sands are also good indicators of past marine conditions.

One of the most important raised beaches in Britain is on the Sussex coastal plain where high sea levels during the Cromerian Complex of interglacials created a chalk cliff 10 kilometres inland and 30 kilometres long. A range of deposits representing full marine through lagoonal to grassland are preserved at the foot of the cliff. The famous site of Boxgrove, discovered as a result of quarrying and described in the previous chapter, is situated on this raised beach (the Goodwood-Slindon Raised Beach).

RIVER SEDIMENTS

River sediments, relating to both extant and former river systems, are those most commonly exploited for sand and gravel extraction. Ancient river channels, known as palaeochannels, are the most important for the preservation and discovery of Ice Age remains. River deposits are typically moderately- to well-sorted and have rounded to sub-rounded gravels.

Of particular importance, when they survive, are sediments that were laid down by low-energy rivers during interglacial periods (Fig. 66; see also Fig. 15, p.18, Phase 3). Fine sands and silts were deposited

Fig. 66: Modern (interglacial) meandering river

Fig. 67: Past interglacial organic river channel (palaeochannel) revealed in an English quarry

Fig. 68: A modern braided river

Fig. 69: Ice Age braided river sediments revealed in an English quarry

primarily in abandoned channel sections or over the wider floodplain during times of flood. They can be recognised as bands of dark, organic-rich silts and sands within or beneath the deposits of sands and gravels (Fig. 67). They are important because they often preserve biological remains such as macroscopic plant remains and pollen, mollusc shells, insect remains, animal bones and possibly also the stone tools of early humans. Such remains permit detailed reconstructions of climate and environment.

Much greater quantities of deposits, generally coarse sands and gravels, were laid down by high-energy rivers under cool climate conditions (Fig. 68; see also Fig. 15, p.18). These much more rarely preserve delicate plant and animal fossils, although the teeth and bones of larger animals may survive, along with stone tools (Fig. 69). Such remains will have been carried by the rivers along with the sands and gravels, and so will usually be distant from their original place of deposition. Their battered and 'rolled' state will bear testament to this transport.

Wind-Blown Sediments

Compared to larger sediments, particles of silt are preferentially picked up and blown along by winds.

Silt was mainly produced by the grinding action of glaciers and was transported in large volumes by glacial rivers. Subsequently these particles were picked up by the wind and re-deposited elsewhere.

Fig. 70: An exposure of loess deposits. Buff coloured primary and secondary loess overlying Late Devensian cryoturbated head, Horton, Gower, South Wales.

Large deposits of wind-blown silts (termed loess) accumulated during the cold periods of the Quaternary (Fig. 70).

Loess is typically a pale yellow colour and small patches of it can be found in Britain. Analysis of loess deposits at sites from Kent to south Devon has revealed a general westward decrease in particle size – indicating that the loess had probably been transported by easterly winds from the large river floodplains that existed in times of low sea levels in the North Sea basin ('Doggerland').

Buried Soils

Fig. 71: Ice-wedge cast formed under arctic conditions and permafrost

Soils form on stable ground surfaces through the action of chemical, biological and physical weathering processes over time. Relict, buried soils may thus be good indicators of past warm interglacial conditions, and their features may be compared to modern soils for information on past climatic conditions. 'Weak' soils can also form in cold climates but the remnants of these soils display evidence documenting the physical action of freeze-thaw cycles (Fig. 71).

Buried soils may sometimes be recognised from the presence of distinct horizontal bands or soil horizons within the deposit (Fig. 72). These are not layers in the sense of different deposits laid down at different times in a sequence. Rather, they are the result of physical and chemical processes going on within the soil. For example, rainwater, which is weakly acidic, brings about the solution and break down of various mineral components in the sediment, particularly calcium carbonate, which may then be lost to the groundwater. This downward process is known as leaching. Conversely, precipitation of dissolved chemicals from the groundwater involves upward movement and gives rise to a variety of deposits such as iron pans (recognisable from their rusty red colour). Thus the horizons visible in buried soils are the result of the movement of weathering products up and down the soil profile. Important diagnostic features of the soil horizons include colour, texture (particle size), and the enrichment or depletion of the calcium carbonate content.

Researchers investigating the sediments at a particular site first systematically describe the sediments, taking into account any structures within them and the particle size, colour, sorting and lithology (or composition) of the sediments, and whether any fossils (e.g. shells) are visible. The data are recorded on a specially designed log sheet (appendix 1). Once the sediments have been described and drawn in detail, inferences can be made as to the forces and agents that acted upon these sediments in the Quaternary past and produced the visible sections.

Fig. 72: A section showing both modern and fossil (buried) soil horizons

Chapter 5: What Remains? – Fossils

Some of the most exciting finds from quarries are plant and animal fossils. These range from large mammal bones to microscopic pollen grains. In this chapter, we introduce these fossils and describe their significance and how they are identified.

Vertebrate Fossils

Finds of complete fossil skeletons are rare and only individual bones and teeth, or fragments of them, are usually found. Only in exceptional circumstances, such as the permafrost zones of Alaska and Sibe-

Fig. 73: The frozen carcass of a baby mammoth, nicknamed 'Dima', found by a bulldozer driver during gold mining operations in Siberia, June 1977. Worth more than gold, he has been insured for $12 million.

Fig. 74: An aurochs (wild cow) depicted on the wall of Lascaux Cave, France. Note the smaller, superimposed figures, including a deer, facing the other way.

ria or the arid caves of Chile and Australia, will the soft tissues, skin and hair of animals be preserved – these can give direct insight into the appearance, diet and health of Ice Age vertebrates (animals with a backbone), as well as preserving ancient DNA (Fig. 73).

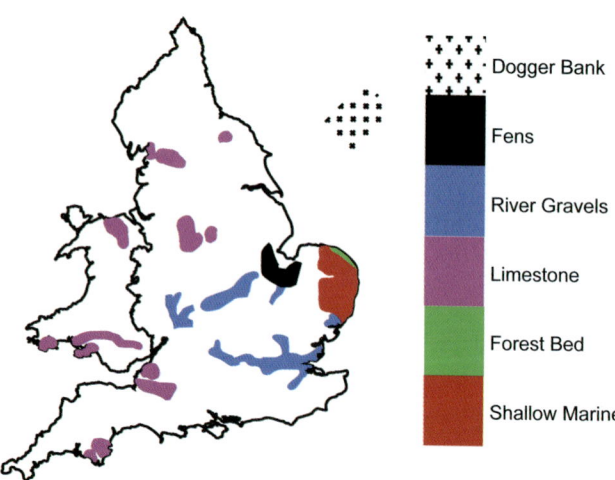

Fig. 75: The main Ice Age mammal fossil localities in Britain

For the last part of the Pleistocene, from around 30,000 BP, some compensation for the extreme rarity of the preservation of soft tissues is the frequent depiction of Ice Age animals in cave art. In naturalistic representations, large mammals are by far the most frequent subject of cave art (representations of humans, by contrast, are rare and usually stylised and schematic). It is clear from the cases where the details can be confirmed that on many occasions these animal representations are highly accurate. The Ice Age hunters had an intense familiarity with the appearance and behaviour of the animals that surrounded them, upon which their lives depended, and this is reflected in their art (Fig. 74).

Fossilised vertebrate remains have been found in a very wide range of environments in Britain, including limestone caves and fissures, former lakeshores and beaches, shallow marine sediments, the Fenland peats and even on the floor of the North Sea (Fig. 75).

However, one of the most common places to find vertebrate fossils is in the fluvial and estuarine deposits laid down by former rivers. Many such finds have been made on the north Norfolk coast in the Cromer Forest-bed Formation, as well as in sand and gravel pits in ancient river deposits in the Thames, Trent, Warwickshire / Worcestershire Avon and many other river valleys. These finds represent both animals that would have lived in the water, such as fish, and other species that may have died on the floodplain and been washed in to the river, or which may have become stuck in mud at the water's edge, been hunted there or even drowned in the water.

EXCAVATION AND SAMPLING

The principal bones of the mammalian skeleton are shown in Figure 76. Most fossils of larger vertebrates found in soft sediments can be excavated by hand using a standard archaeological trowel, switching to smaller wooden or plastic implements when near the bone surface so as not to damage it (Fig. 77). The specimen should preferably be recorded in three dimensions, tied into a site plan. As a minimum, however, recording should include descriptions of the deposit in which the specimen was found, a photograph or sketch of its position in the ground or section, and an OS grid reference for the site.

Specimens can be cleaned with warm water and a soft brush prior to being allowed to dry slowly (never in direct sunlight) but more fragile remains may require consolidation *in situ* (e.g. using a plaster 'jacket') and lifting as a block for later excavation in the laboratory. Full records of the conservation measures taken should be kept and specialist advice sought, especially prior to the application of any glues or other 'consolidants', as these can affect later analyses.

For smaller vertebrate remains, which cannot easily be seen in the field with the naked eye, bulk sediment samples (a minimum of 10 litres/10-20kg) should be taken for wet-sieving. Bulk samples are normally taken as a column of separate samples through the deposits, from the base to the top of a section, so that any change through the sequence can be identified. All sediment should be wet-sieved

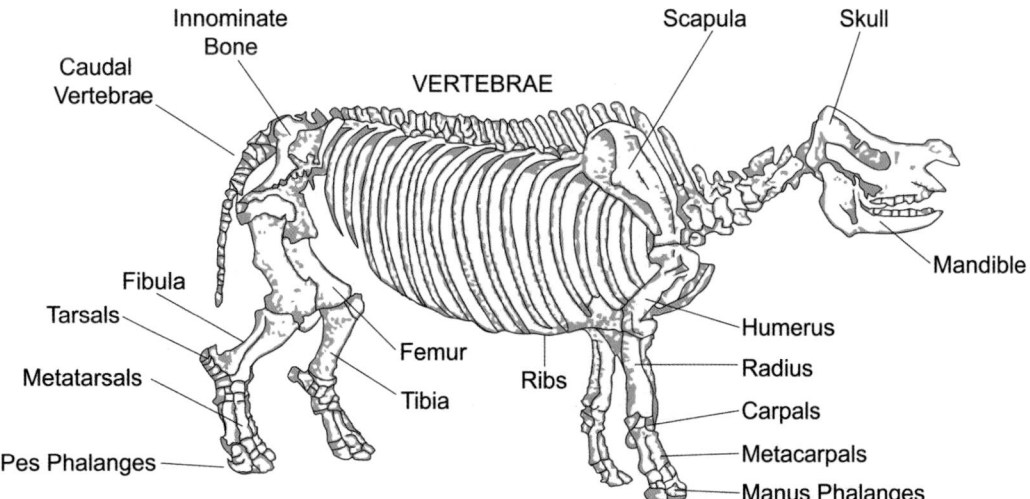

Fig. 76: The skeleton of a rhinoceros, showing the names of some of the principal bones of the mammalian skeleton

Fig. 77: Excavating the remains of a fossil aurochs (wild cow)

using a 0.5mm mesh. When the residue is clean and dry it can then be sorted under a low-power (10x) binocular microscope for specimens.

Once the fossil bones have been recovered, the first stage of analysis is to identify the part of the body represented, then the species, before establishing age and sex, as well as the number of individuals present. Identification of remains is carried out using illustrations and reference material.

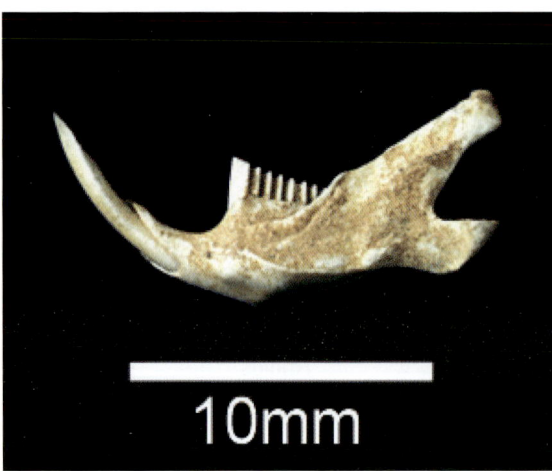

Teeth are particularly important for the identification of fossil mammals. Teeth are the toughest part of the skeleton and therefore the part that often survives best. The teeth of elephants (including mammoths) are both big and strong, and isolated elephant teeth are one of the most common fossil finds in sand and gravel quarries. Furthermore, the teeth of an animal are closely adapted to its diet. Teeth thus provide one of the best means of identifying fossil remains to species, as is shown in the brief outline below of the main fossil vertebrates encountered.

Fig. 78: Vole jaw, showing the long curving incisors (front teeth) and 'radiator-like' molars (cheek teeth) typical of rodents

SMALL VERTEBRATES

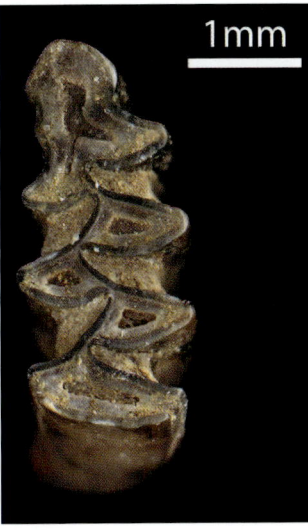

Fig. 79: Side view and biting-surface view of vole molar

Fish remains commonly include vertebrae (the backbone), fin spines, teeth and, occasionally, the fragile scales. Amphibians such as frogs and toads are characterised by their vertebrae and 'fluted', hollow long bones, and reptiles by their vertebrae, although tortoise shell fragments may also be found. Bird remains are rare, particularly on open sites, and are light and hollow; the beak and limb bones being most recognisable.

Most small mammals can be identified to species level by their teeth, particularly the rodents (e.g. mice, voles, lemmings, squirrels and beavers), which possess two pairs of long, curving enamel incisors (front teeth) and 3–4 molars (cheek teeth) in each jaw (Fig. 78). Vole and lemming molars resemble small radiators from the side and consist of a complex series of interlocking enamel triangles when viewed from the biting surface (Fig. 79), whereas mice have low-crowned, rounded molars. Insectivores such as hedgehogs, shrews and moles have a long row of sharply pointed teeth (Fig. 80).

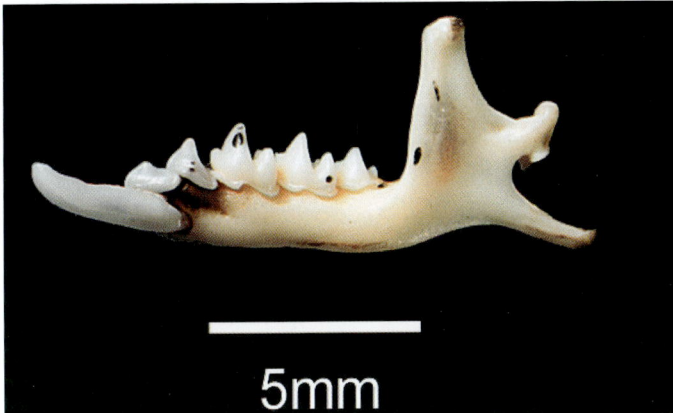

Fig. 80: Shrew jaw, showing the long row of sharply pointed teeth typical of insectivores

CARNIVORES

A range of fossils of large carnivores occurs in British Ice Age terrestrial deposits, particularly in caves. They include cave and brown bears, lions, sabre-toothed cats, leopards, wolves and spotted hyaenas. Smaller predators such as foxes, wild cats and mustelids (otters, weasels and their relatives) are also found.

The carnivores are readily identified by their prominent canine teeth (in sabre-toothed cats these evolved into dagger-like points with serrated edges) and by the presence of carnassial teeth. The carnassial teeth comprise the first lower molar and the fourth upper premolar and have a blade-like structure for shearing meat. Depending on the dietary adaptation of the carnivore in question, a range of slicing, crushing or grinding teeth will also be present (Figs 81–83). Bears differ from the other carnivores in that they lack carnassial teeth and have only low-crowned cheek teeth, reflecting their omnivorous diet (Fig. 84).

Ice Age

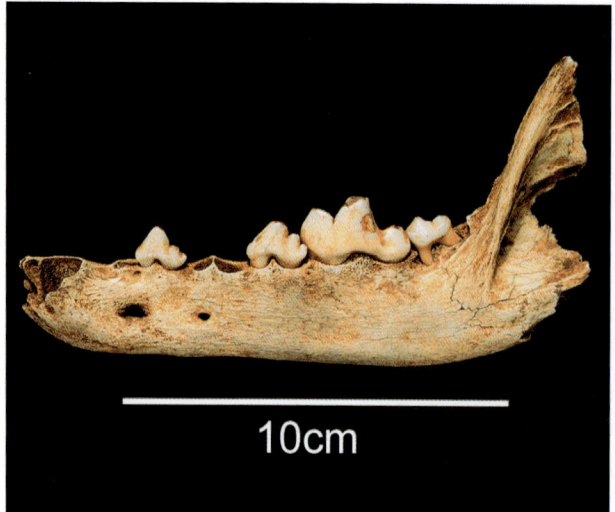

Fig. 81: The mandible (lower jaw) of a wolf

Deer and Bovids

Deer and large bovids (bison and aurochs - wild cattle) are some of the most abundant Ice Age fossils. The species encountered range from reindeer and musk ox, which were restricted to cold climate episodes, to fallow deer and aurochs in interglacials. Bison, red deer and giant deer are common to both warm and cold climates, although the giant deer occupied only open habitats on account of their enormous antler span (up to 3m).

Both deer and bovids have typical herbivore teeth, consisting of linked crescent-shaped molars with enamel infoldings. In deer, the antlers are most diagnostic, in particular the surface texture, the arrangement of tines (projections off the main beam) and the degree of palmation or flattening of the end (Fig. 85). The

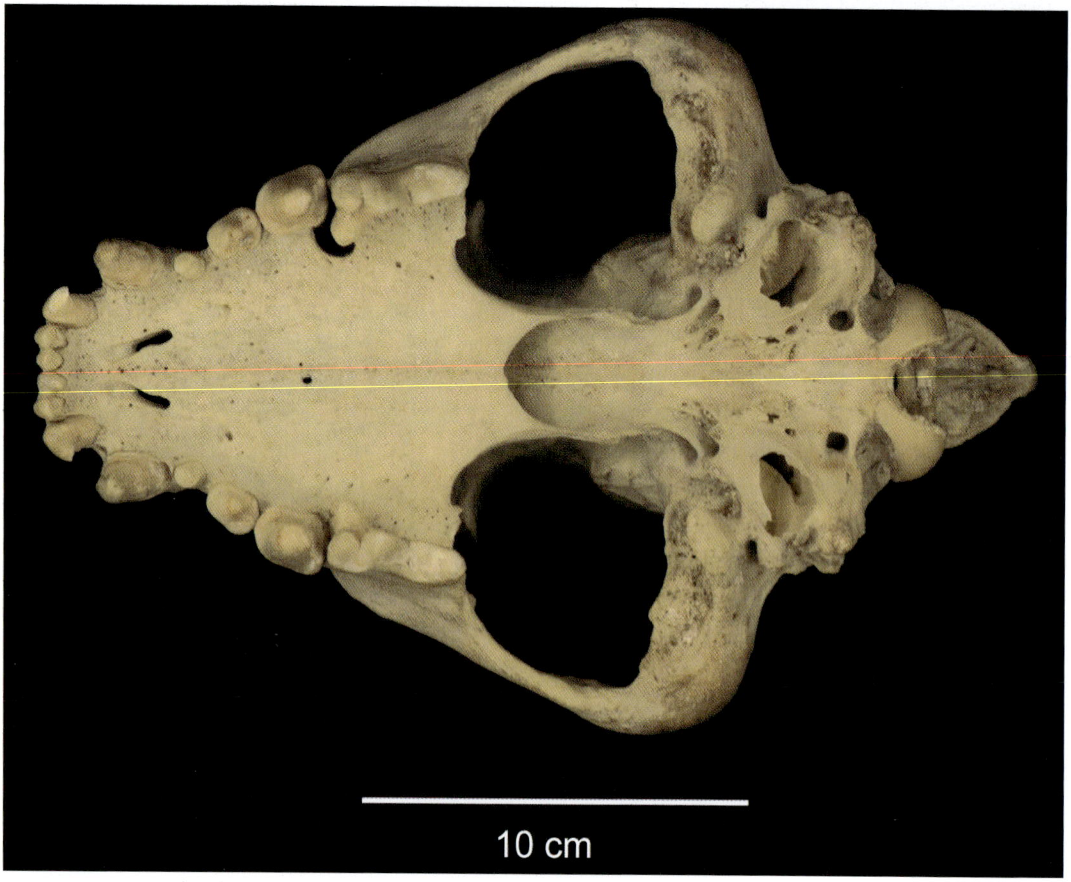

Fig. 82: The skull of a hyaena from beneath, showing dentition in the upper jaw

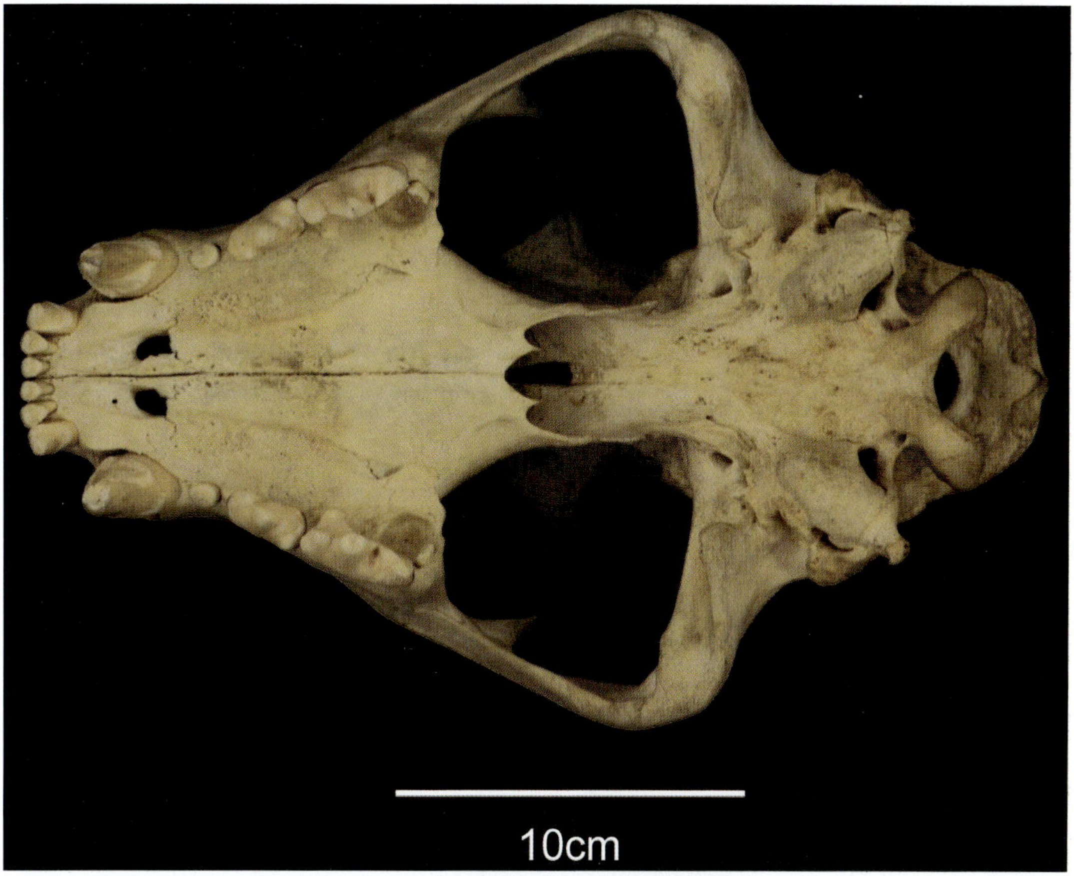

WHAT REMAINS? – FOSSILS

Fig. 83: The skull of a lion from beneath, showing dentition in the upper jaw

Fig. 84: The upper jaw of a bear – the low-crowned molars (cheek teeth) reflect an omnivorous diet and contrast with the slashing teeth of carnivores with a greater meat component in the diet, such as lion

Fig. 85: Red deer antler

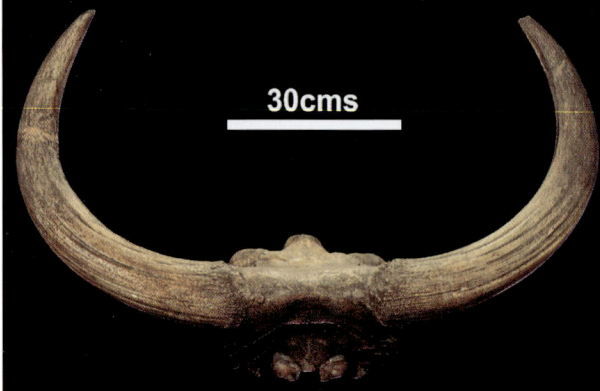

Fig. 86: Bison horns

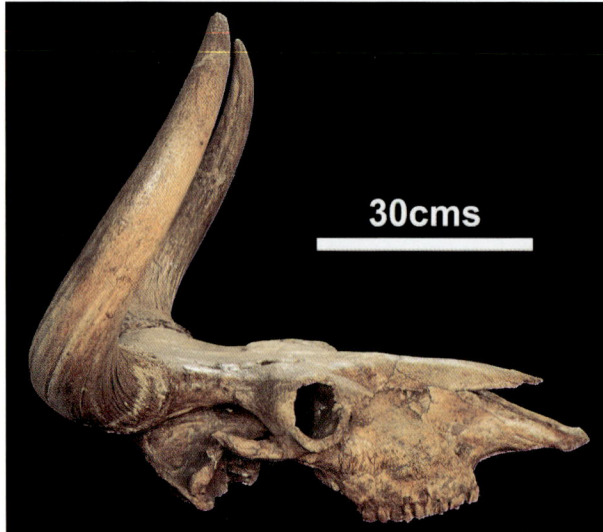

Fig. 87: The skull of an aurochs (wild cow)

shape of the horns in bovids is also characteristic – the single upward tilt in bison (Fig. 86), the upward and forward projection in aurochs (Fig. 87) and the 'helmeted' form of the musk ox. The limb bones are similar in both, although those of the deer are relatively longer and more slender.

Elephants

Elephants (including mammoths) were the largest of the Ice Age megafauna (all animals weighing more than about 40kg) in Europe. The straight-tusked elephant (*Palaeoloxodon antiquus*) was found exclusively under temperate (warm) conditions whereas the mammoth (*Mammuthus*) lineage was present in both warm and cold episodes. The ivory tusks of these animals are immediately recognisable – long and straight in *Palaeoloxodon* and downward-spiralling in *Mammuthus*.

Elephants have only four teeth in their jaws at any one time and go through a succession of six sets in their lifetime. The teeth consist of large enamel plates or 'lamellae' stacked one behind the other; where the enamel is worn away at the surface, it creates a characteristic pattern of diamond shapes in *Palaeoloxodon* (Fig. 88) and parallel strips in *Mammuthus* (Fig. 89). Over the 2.6 million years of the Ice Age, as mammoths moved from temperate habitats, where they ate soft vegetation, into cold steppes, their molars doubled in height and in the number of enamel plates present, in order to cope with a diet of abrasive grasses (Fig. 90).

Rhinos and Horses

Rhinos and horses are members of the mammalian order Perissodactyla, meaning that they have an odd number of toes (3 in rhino and 1 in horse). The limb bones and toes of these species are therefore very diagnostic, in particular the cannon

Fig. 88: Straight-tusked elephant lower tooth

Fig. 89: Mammoth upper tooth

bone (the 3rd metacarpal or metatarsal – see Fig. 76 above) of horses and the digits of both groups. Several species of rhino are known from Ice Age interglacials, adapted to either woodland or grassland habitats, whereas the woolly rhino is a characteristic component of cold stage faunas. Horses are common to both climatic phases.

Rhinos have large, robust teeth with a complex pattern of ridges and thick enamel. The upper molars are square-ish, whereas the lower molars are 'w' shaped. The woolly rhino has an additional isolated enamel ring in the upper molars that distinguishes it from interglacial species (Fig. 91). Horses possess tall, column-like teeth with complex enamel folds in both the upper and lower sets (Fig. 92).

Vertebrates and Interactions With Stone Age People

Vertebrate remains, particularly mammals, sometimes show evidence of modification by early humans, such as cutmarks left by stone tools (Fig. 93). These can be distinguished from scratches and other natural marks by their position, as they are usually arranged as parallel incisions near the joints or major muscle blocks, and by their sharp, v-shaped cross-section under the microscope. Other bones may have a characteristic 'spiral fracture' where the fresh bone was broken and twisted apart for marrow extraction (Fig. 94). As well as supplying meat, fat and marrow, which were important in the diet, animals also provided a source of furs, fuel (fat or dung could be burnt by hominins) and raw materials for making tools or art objects (Fig. 95).

Interpreting Vertebrate Fossils

Many Ice Age vertebrates or their close relatives are still alive today. Studying fossil remains can shed light on past environments (by examining anatomical adaptations and modern habitat preferences), as well as revealing how many species (particularly mammals and birds) changed their size and shape in response to Ice Age climate change. Many mammal lineages show clear patterns of evolution and extinction, in addition to local patterns of presence and absence, which may be used to construct faunal histories. As with all palaeontological studies, an understanding of the taphonomy (i.e. how the bone assemblage formed) is critical.

As was noted in chapter 2, these faunal histories provide an important dating technique because different characteristic assemblages of mammals were present during different periods of the Ice Age. This dating technique, known as biostratigraphy, is described in more detail in chapter 7.

In favourable circumstances it is possible to study the evolutionary relationships of extinct mammals

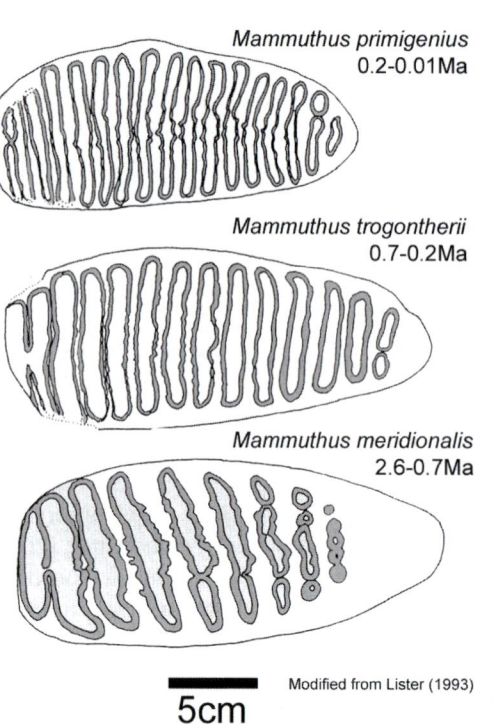

Fig. 90: The increasing number of enamel lamellae (plates) forming the teeth of different mammoth species through time

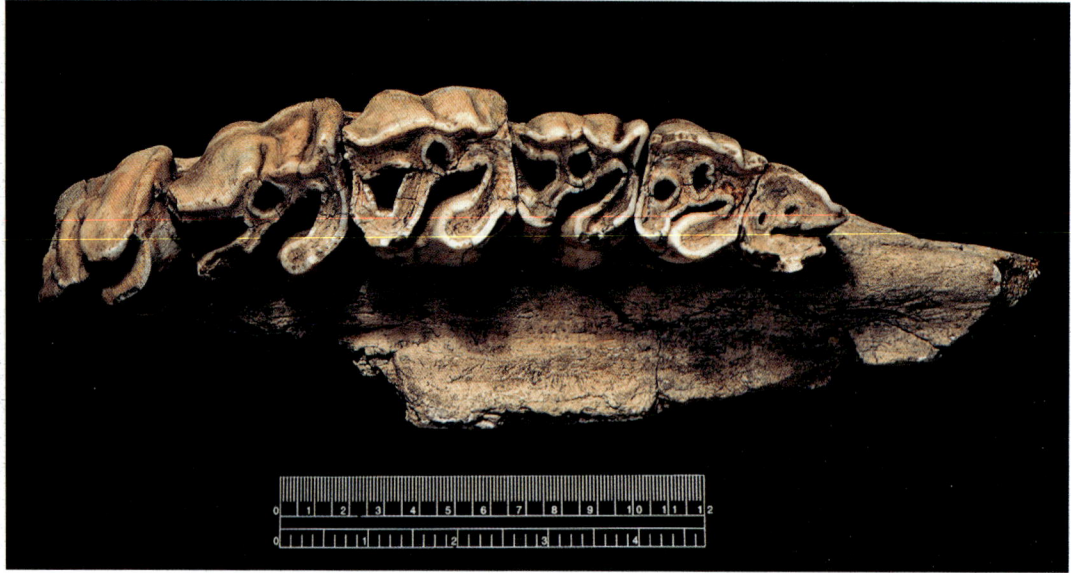

Fig. 91: Woolly rhinoceros upper teeth

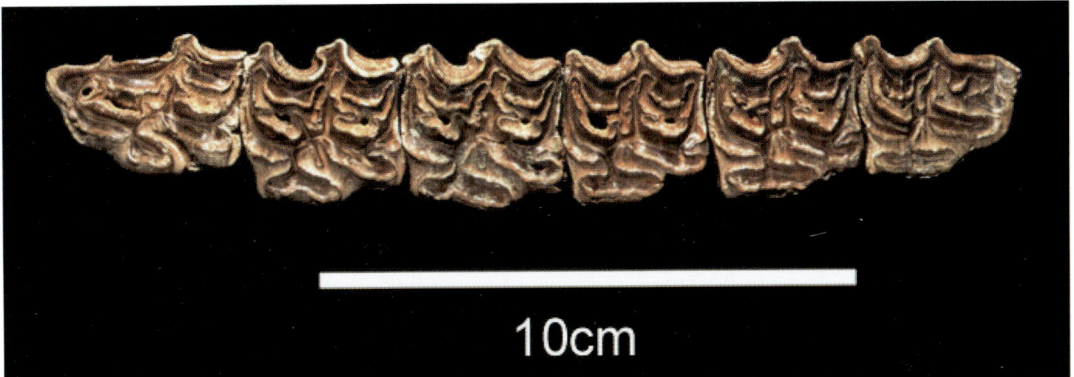

Fig. 92: Horse upper teeth

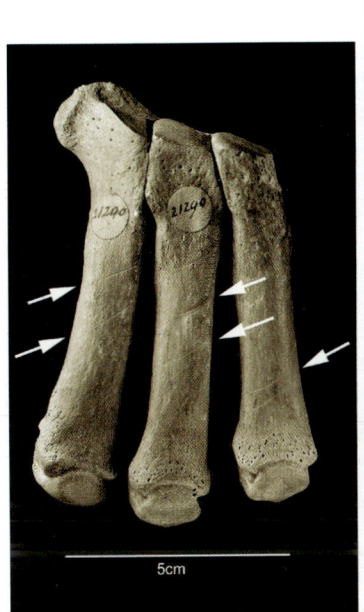
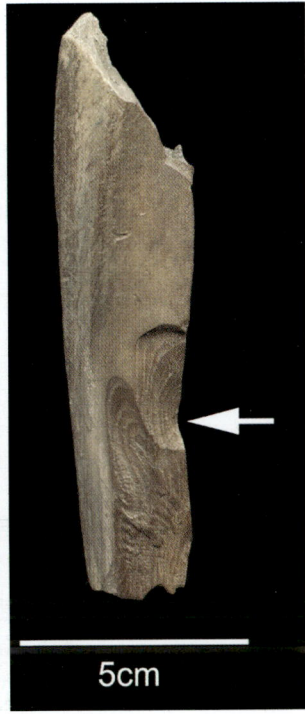

Fig. 93: Cut-marked brown bear paw bones (arrows indicate cuts)

Fig. 94: Reindeer bone broken for marrow (arrow indicates impact)

Fig. 95: Carved mammoth figure

such as mammoths and the woolly rhinoceros through the extraction of ancient DNA from their bones. The ancient DNA sequences can be compared with the DNA extracted from other fossils as well as with the DNA of living species.

Another relatively new scientific technique used in the study of fossil bones is stable isotope analysis. Stable isotope analysis was encountered in chapter 2, where the relative proportions of different stable isotopes (variant forms of the atom) of oxygen in the shells of tiny marine organisms was seen to reflect the chemical composition of the sea water in which the organisms lived. From this, inferences about past climates and climate change could be made. The same principal applies to other animals, and analysis of the varying proportions of the stable isotopes of oxygen, nitrogen and carbon preserved in bone can shed light on past climate as well as on diet, in particular the relative importance of meat, plant resources and marine foods to both animals and humans.

Hominin Fossils

As was seen in chapter 3, the fossil remains of ancient humans (hominins) pre-dating *Homo sapiens* have only very rarely been found in Britain. The list (excluding the Piltdown forgery!) is very short: a shin bone and two teeth from Boxgrove, probably of *Homo heidelbergensis*, dating to c.500,000 BP; skull fragments from the Thames gravels at Swanscombe, perhaps transitional between *Homo heidelbergensis* and *neanderthalensis*, dating to c.400,000 BP; a collection of seventeen teeth from Pontnewydd Cave in Wales, representing early Neanderthals, dating to c.225,000 BP; and another small collection of Neanderthal teeth from La Cotte de St Brelade, Jersey, dating to c.150,000 BP (at this time Jersey was not an island but an area of high ground on the vast plain that stretched from France towards England). All the pre-Neanderthal finds – that is the fossils from Boxgrove and Swanscombe – were uncovered at quarry sites.

The extreme rarity of fossil hominin remains from Britain no doubt in part reflects the rarity of the hominins themselves. Nevertheless, for some periods and places, for example the Hoxnian interglacial (c.400,000 BP) in southeast England, stone tools (notably the easily recognisable handaxes) have been recovered from quarries in quantities that measure in the hundreds of thousands. Hominin remains might well have been present but were possibly not recognised and recovered. The extreme rarity and scientific significance of finds of early hominin remains renders any such find, no matter how fragmentary, of the highest importance.

Human burials, when the skeleton is relatively intact, are easy enough to recognise. However it was not until the time of the 'classic' Neanderthals, that is within the last 100,000 years, that deliberate burial of the dead in a grave of some sort occurred with any frequency (and even so it remained rare). Deliberate burial in a grave greatly enhances the chances of a human skeleton surviving relatively intact, potentially preserving it from the depredations of scavengers and the natural processes of break up and dispersal caused by water and wind. A cave environment can provide some protection from some of these processes, and all known Neanderthal burials have been found in caves.

Up until the time of the later Neanderthals hominins do not seem to have buried their dead or disposed of their remains in any formal manner, although 400,000-year-old collection of human remains from the Sima de los Huesos ('Pit of Bones') in Spain raises some intriguing questions. Apparently the body of a dead hominin suffered the same fate as the carcass of any dead animal, and various natural processes led to the disintegration and dispersal of the bones. Like the skeleton of any other animal, therefore, the preservation of an intact hominin skeleton and its discovery will be vanishingly rare and limited to highly improbable circumstances such as rapid freezing and natural burial. In almost all cases, as with other animals, all that will survive, if anything survives at all, is scattered fragments, mainly of the more robust parts, such as teeth, long bones and skull fragments. This explains the character of the remains from the few British sites (listed above) that have produced hominin remains.

Although they have only very seldom been recognised and recovered, one of the most promising locations for the finding of early hominin fossils is in sand and gravel quarries. They can potentially be found in any deposit that has produced the bones of other large mammals together with the stone tools the hominins made – a not infrequent occurrence. More awareness of the importance of these deposits, and of the need to report and recover the fossils uncovered, could undoubtedly lead to new discoveries of early hominin remains. The key ingredients for success are knowledge and enthusiasm.

WHAT REMAINS? – FOSSILS

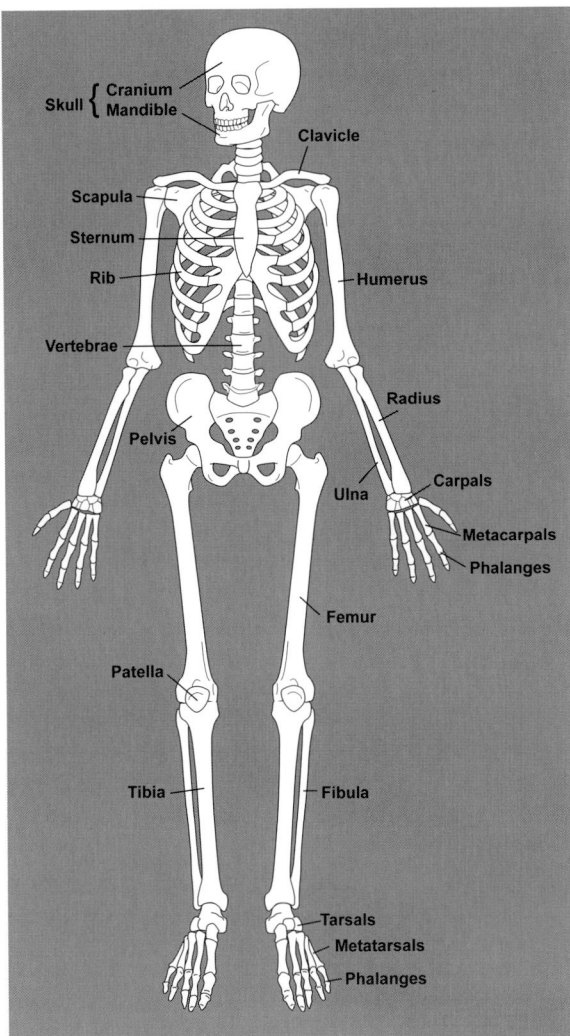

Fig. 96: The human skeleton showing the principal bones and bone groups

Figure 96 provides a diagram of the (modern) human skeleton. As described in chapter 3, earlier hominins differ mainly in the shape of the skull and the robustness of the body. We have emphasised that finding a complete skeleton of an early hominin is vanishingly unlikely, and indeed the finding of a skeleton buried amongst sand and gravel deposits is almost certainly an indication that the burial belongs to a grave dug relatively recently, i.e. within the last 6,000 years or so. More than one scholar has been mislead into attributing a burial dug down into Ice Age deposits as belonging to the Ice Age deposits themselves. Nevertheless, these not infrequent later burials will almost always be of archaeological importance (and all finds of human remains should be reported to the Home Office).

INVERTEBRATE FOSSILS

The remains of invertebrates (animals without a backbone) are amongst the most commonly occurring in Ice Age deposits. They include molluscs (e.g. snails and mussels), beetles and other insect remains. During the Quaternary, relatively few new species have evolved, and very few have become extinct. Thus Ice Age invertebrate species can be used to infer past environments because they still have living descendants whose habitat preferences are well known and often very specific. The importance of these animals for the reconstruction of past environmental and climate change is thus entirely disproportionate to their small size. Without knowing this, it is easy to overlook deposits uncovered in quarries and elsewhere that appear insignificant but are potentially of considerable scientific importance.

Beetle fossils can be used to reconstruct past temperatures and the calcium carbonate content of mollusc shells can be used to reconstruct ancient water temperatures (from the oxygen isotopes in the shells). The age of the shells, and therefore the age of the deposits, can also be determined by the amino-acid or radiocarbon content of the shells (see chapter 7).

The size of invertebrate fossils ranges from under 1 millimetre to 20 centimetres. They can be seen with the naked eye when present in large numbers, but if present in lesser quantities (or if only smaller species are represented) then they can only be detected when a bulk sediment sample has been sieved in the laboratory.

Invertebrate fossils tend to be best preserved in fine-grained sediments such as silts or clays. These fossils are quite fragile and if incorporated into gravels they are often smashed and destroyed.

Ice Age

Molluscs

The different species of molluscs occupy a wide range of habitats from water bodies of almost any kind to grassland, woodland and scrubland. Many species are restricted to very particular habitats. Arctic/alpine species such as *Columella columella* only live in unshaded places, in tundra or mountains for instance. By contrast *Clausilia pumila* is found in old woods in central Europe. Although neither of these species is found in Britain today, both have been found as Ice Age fossils indicating the climate and environment of Britain at that time (Fig. 97).

The mollusc fauna of Britain today consists of about 220 species, although not all of these have been found as Ice Age fossils. Most Ice Age snails still have no English name and are known only by their Latin names.

Mollusc Types – Gastropods

Gastropods (snails) have coiled or cone-shaped shells and live on land and in water. They move by contractions along a muscular 'foot' on the underside of the animal, and can retract their soft parts into the shell to avoid danger or drought. Figure 98 shows some common gastropod forms.

Mollusc Types – Bivalves

Animals such as mussels and clams are known as bivalves (Fig. 99). Their shells are formed of two halves (valves) joined by a hinge which is held together by an elastic ligament. The animal can open its shell to feed, breath and reproduce but can also hold the shell closed as a defence against predators or

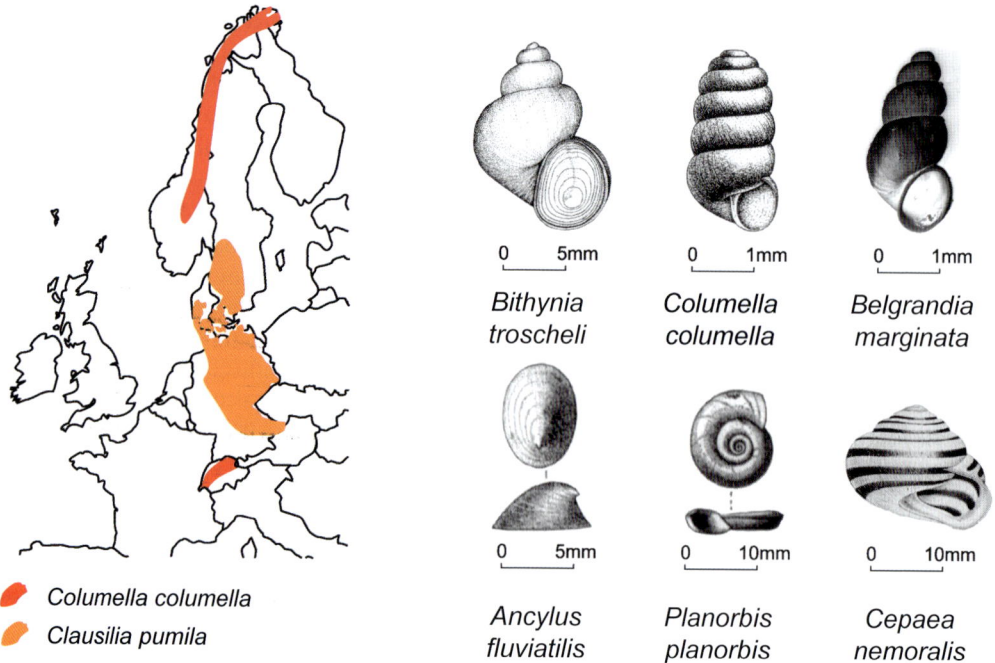

Fig. 97: Modern distribution of two snail species, *Columella collumella* and *Clausilia pumila*, which lived in Ice Age Britain

Fig. 98: Some examples of common gastropod (snail) species

WHAT REMAINS? – FOSSILS

Fig. 99: Some examples of bivalve (mussel and clam) species

fast moving water. Bivalves live only in water where they can either burrow into the bottom sediments or attach themselves to stones or other hard parts of river beds.

FINDING AND INTERPRETING MOLLUSCS

Fig. 100: Top: Shells exposed in the sands of the Thames. Bottom: The shells after sieving.

Molluscs are usually found in sand or mud rather than in gravel or clays (Fig. 100). Silted-up Pleistocene channels that have been exposed by gravel quarrying often contain the most shells – these are mainly freshwater shells, but the channels may also contain some land shells washed into the river by floods.

Shells are often visible because they are bleached white, although the colours of the shells in life may also be preserved. Deposits from warm interglacial periods usually have a wide variety of species (70 or 80), while those from cold climate conditions tend to have fewer species (10 or so), but many examples of each species may be present. Because many shells are small in size (under 5 mm commonly), only the largest species are easily visible in quarry faces, although faces recently wetted by rain may reveal shells that have been washed free of their enclosing sediment. Shells can also be found in more unusual settings, including hollows in fossil animal bones and skulls that they have colonised after the death and decomposition of the host animal (Fig. 101).

If shell beds are found, then the best way to investigate them is to take bulk samples (e.g. 20 litres), noting the depth and location of the sediments. These samples are often taken as column samples, from the base to the top

79

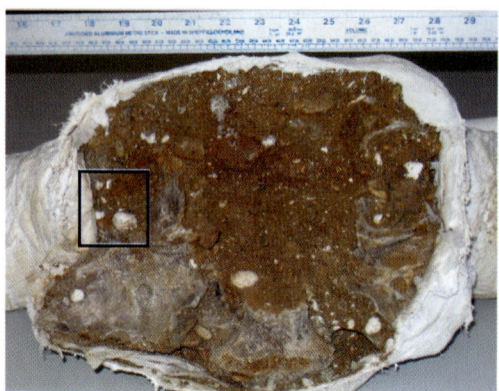

Fig. 101: *Corbicula* shells within an aurochs' (wild cattle) skull

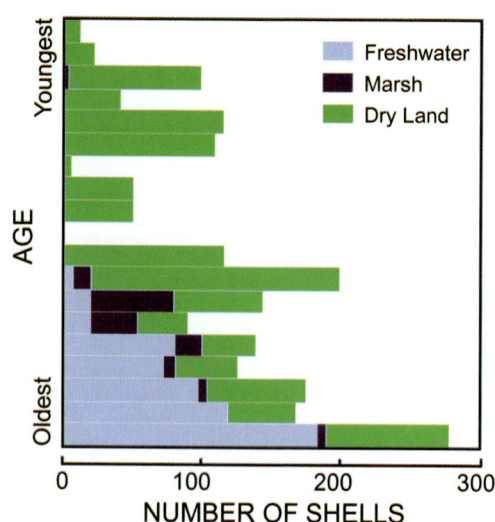

Fig. 102: Mollusc evidence for changing environments 200,000 years ago in Buckinghamshire

of a section, in order to identify any changes through the sequence. In the laboratory the sample is sieved and the shells removed for identification.

The 'specialisations' of different mollusc species allow researchers to use fossil assemblages (groups of individual fossils) to produce detailed reconstructions of Ice Age environments. For example, at Marksworth in Buckinghamshire, analysis of the fossil mollusc assemblages based on modern ecologies shows the transformation of the site from freshwater pond to dry land 200,000 years ago (Fig. 102).

BEETLES

About a quarter of all known species are beetles (Coleoptera). Different beetle species occupy almost all habitats from freshwater bodies to scrubland and woods. Most beetle species are specialists, adapted to particular environments and conditions such as narrow temperature ranges. For example, *Stephanocleonus eruditus* (Fig. 103) is a weevil that is only found near snowfields in the alpine tundra of Siberia. Some water beetles are entirely dependent on running streams while others inhabit only still water. Other beetle species have specific dietary requirements. There are carnivores, reed-eaters, tree-eaters and dung feeders, beetles which eat only fresh carcasses and others which eat only dry carcasses.

Beetles hardly evolved at all during the Quaternary, and the same associations of beetles are often found together both in fossil assemblages and at the present day. This fact, in combination with the very specific habitat requirements of many beetles, means that beetle remains can be used to undertake detailed environmental reconstruction.

The beetle fauna of Britain consists of over 3,800 named species, although not all of these have been found in Ice Age deposits. Conversely, many species of beetle are found in Ice Age deposits that are now foreign to Britain, including species found today only in arctic or alpine regions. Like snails, most beetles have no English name and are known only by their Latin names.

Fig. 103: Head capsule of *Stephanocleonus eruditus*, a species of weevil

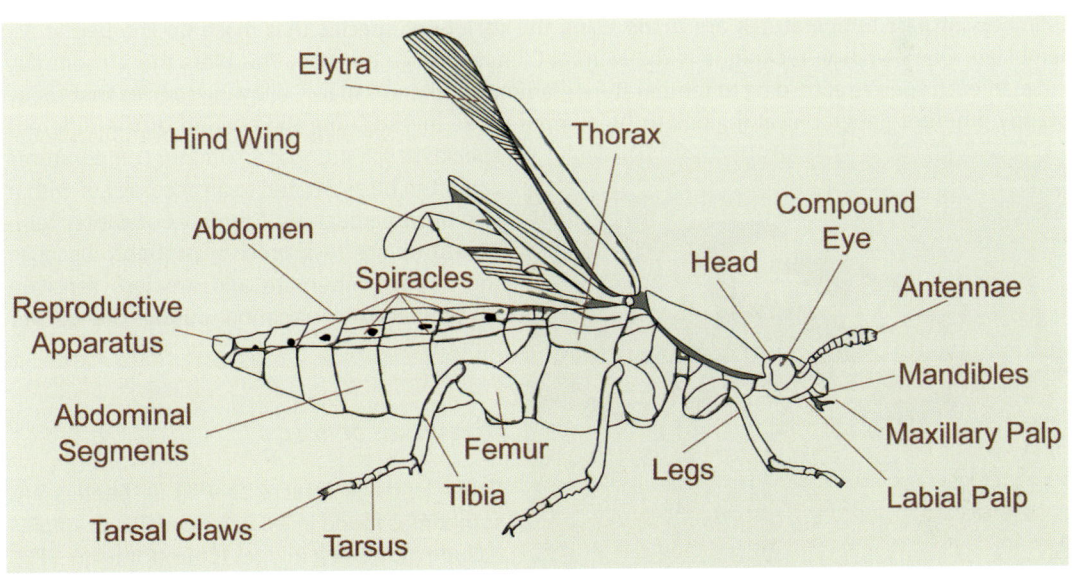

Fig. 104: Diagram of a beetle showing the principal body parts

Finding Beetles

Fig. 105: Beetle wing cases

Beetles tend not to survive intact in Ice Age deposits. Generally what survives are the wing cases (elytra) and the thorax (Fig. 104). However the distinctive features of the wing cases (Fig. 105) and thoraxes often enable beetle remains to be identified to species level.

Beetles are best preserved in fine-grained sediments such as silts, sands and organic clay deposits within bodies of gravel. They are rarely visible to the casual observer and so bulk sediment samples (generally 20 litres) are taken for sieving in the laboratory. Again, it is important to accurately record the stratigraphic context and location of the sediment samples.

Climate Reconstruction Using Beetles

Due to the climatic sensitivity of many species, fossil beetle remains are of particular importance for the reconstruction of past temperatures. While the oxygen isotope record from deep sea cores and ice cores can be used to reconstruct global patterns of climate change, beetle remains can be used to reconstruct summer and winter temperatures at the local and regional scale.

Estimates of past temperatures are made using the variety of species that make up the beetle assemblage. One common technique is the Mutual Climatic Range method. This plots the present-day range of each species according to temperature gradient or 'climate space', allowing species that today occupy different geographical regions to be compared. By investigating the overlap of the different species in climate space, summer temperatures can often be estimated to an accuracy of within two or three degrees. Estimates of winter temperatures are less precise, probably because the beetles hibernate and are less sensitive to temperature variation during the winter months.

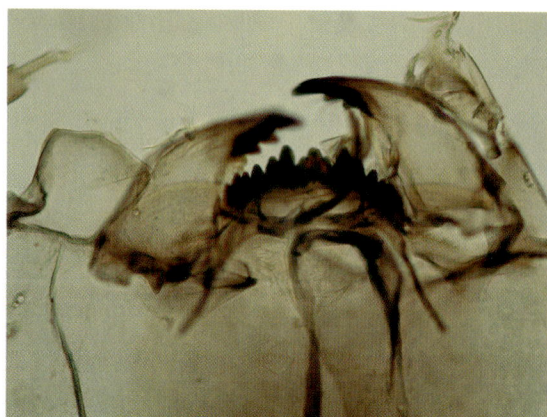

Fig. 106: Chironomid head capsule x1000

Other Kinds of Insects

Other kinds of insects as well as beetles are sometimes found in Ice Age deposits. Examples include non-biting midges (Chironomidae – Fig. 106) and caddis flies (Tricoptera). These can also be used for environmental and climate reconstruction although they are not as well studied in this respect as beetles.

ANCIENT VEGETATION

As well as animal fossils, sands and gravel quarries can also uncover deposits containing plant remains. These provide direct evidence of the vegetation growing in and around the deposits in which they are found. Plant remains can be divided into macrofossils, generally visible to the naked eye, and microfossils, which need to be studied under the microscope.

PLANT MACROFOSSILS

Fig. 107: Fossil pine tree stumps buried in peat in Scotland

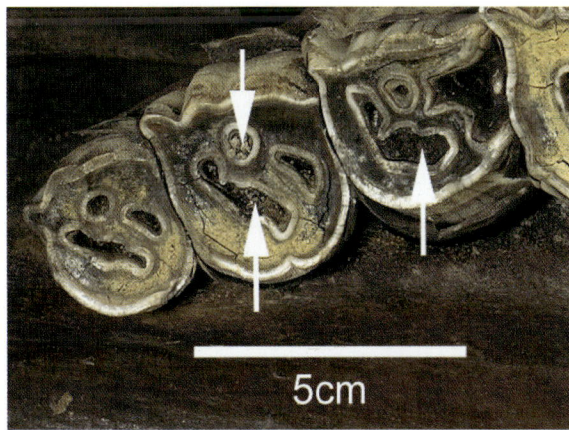

Fig. 108: Plant fragments embedded in the teeth of the Whitemoor Haye woolly rhino

The term plant macrofossils covers all larger fossil plant parts from tree trunks (Fig. 107) to seeds, leaf fragments, stems and roots. Like invertebrate animals, plants have also changed very little through the Quaternary. Plant remains are best preserved in fine-grained river (fluvial) deposits, lake sediments, and within organic and peat deposits.

The particular advantage of plant macrofossils over pollen (see below) is that they are frequently recognisable to species level, so precise information can be obtained about past environments. Also, as they are relatively large and tend to travel only short distances, the presence of plant macrofossils generally provides good evidence that a specific suite

Fig. 109: A sedge seed magnified 40x

of plants grew in the immediate vicinity. They are therefore reliable indicators of how local environmental conditions changed over time.

Plant fossils are also sometimes found in direct association with animal remains, giving us important insights into the diets of the animals. For example, plant remains embedded in woolly rhino teeth from Whitemoor Haye quarry, Staffordshire, indicate a diet of grasses and herbs (Fig. 108).

Although plant macrofossils are usually visible to the naked eye, in order to collect the smaller remains such as seeds (Fig. 109) and leaf fragments, and to ensure that a representative assemblage is studied, they are generally collected in bulk sediment samples (usually 20 litres). These samples are wet-sieved using a fine (125μm) mesh to collect the residue containing the remains. Once the residue is dry it is sorted under a low-power (10x) binocular microscope for specimens. The specimens are identified using keys, drawings and reference collections.

PLANT MICROFOSSILS – POLLEN AND SPORES

Pollen (Figs 110 and 111) is produced by the male parts of flowering, seed-producing plants. It contains the male gamete (germ cell) that can fertilize the female ovule, which then becomes a seed. The amount of pollen produced by a plant will depend on how the plant is pollinated. Plants such as roses, which rely on insects to transfer their pollen to other plants, produce less pollen than plants that are wind pollinated, such as grasses.

Spores are produced by 'lower' plants with no true flowers or seeds (e.g. ferns) and are entirely wind

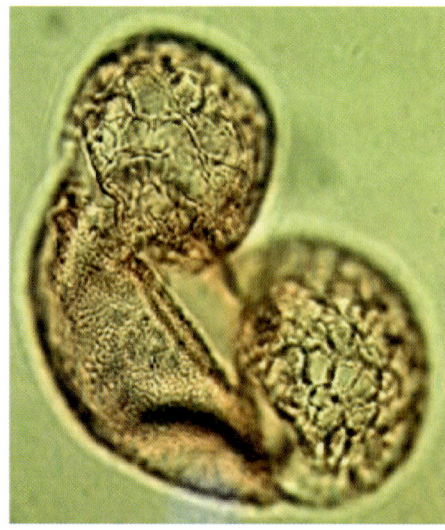

Fig. 110: Pine pollen grain 1000x

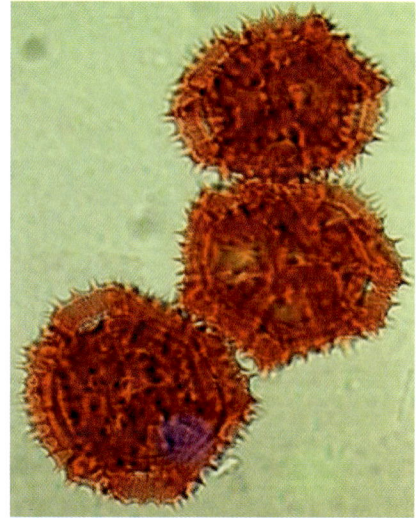

Fig. 111: Dandelion pollen grains 1000x

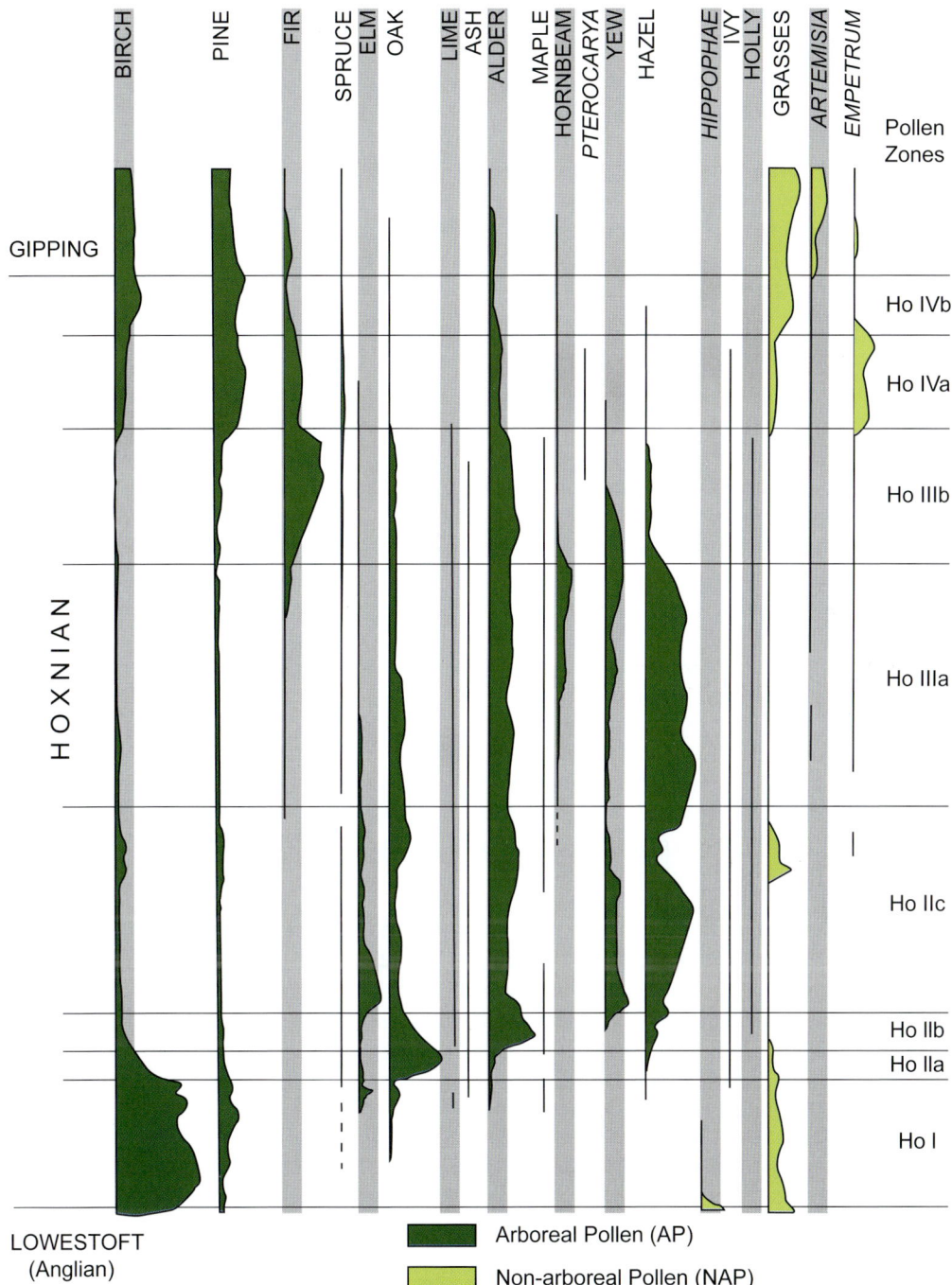

Fig. 112: Pollen diagram from Marks Tey in Essex showing landscape development during the Hoxnian interglacial (about 427,000–364,000 BP). The proportions of the pollen of the different types of plants changes from the beginning of the interglacial (bottom) to its end (top) as mixed woodland develops and then disappears again. The sequence is divided into *pollen zones*.

dispersed. As pollen grains and spores are so small they often find their way into river systems and lakes where they are incorporated into the sediments. Despite not being readily visible to the naked eye, many Ice Age sediments contain and preserve pollen grains.

All pollen grains are microscopic with few exceeding 1/10th of a millimetre in size. Unlike plant macro-fossils, they can generally only be identified to genus or family level (e.g. *Betula* = birch). Only rarely can pollen grains be identified to species level.

Pollen grains can provide valuable insights into regional vegetation cover and climatic conditions. For example, at Marks Tey in Essex pollen evidence (Fig. 112) from the Hoxnian interglacial (approximately 400,000 years ago) shows the landscape developing from open grassland with birch and pine (as conditions became warmer) to deciduous woodland (full interglacial conditions) and back to open grassland again (moving back into glacial conditions).

Finding and Studying Pollen

Pollen grains are best preserved in organic sediments (e.g. peat, some silts and clays), lake or pond deposits. The pollen raining down on the bed of a lake or on a peat bog, for example, will accumulate as the deposit accumulates. Thus the changing types and different proportions of pollen at different levels in the deposit, from bottom to top, will reflect the changing environment through time. This important property is often exploited in the way that pollen samples are taken. For example column samples may be taken through the sediment exposed in a section, often using monolith tins (Fig. 113), so that the stratigraphy of the sediment is preserved. The column samples are wrapped in cling film (to protect the sediment) and the top and bottom of the column is marked. The location of the sample and its stratigraphic position are carefully recorded. Columns of sediment containing pollen may also be extracted using various coring techniques.

In the laboratory the pollen is extracted using chemical treatments and studied under a high-power microscope. Where a column sample has been taken, samples of the pollen will be extracted from different levels up the column so that change through time can be studied. The pollen grains are identified and the different types of pollen grain are counted so that the relative abundance of different types of plants in the environment may be estimated.

Because different types of plants produce different quantities of pollen dispersed in different ways, pollen analysis is not straight forward. Some of the pollen will come from plants growing near to the deposit in which the pollen accumulated but other pollen will have come from much further away. The results of pollen analysis are shown in the form of a *pollen diagram*, with curves showing the relative abundance of different types through time (Fig. 112).

Pollen analysis plays a very important role in reconstructing climate and environmental change through time. Before the advent of the oxygen isotope records from deep-sea and ice cores (see chapter 2), pollen analysis was the principal method used to reconstruct climate change through the later stages of the Pleistocene and, along with beetles, it remains fundamentally important in the reconstruction of local and regional patterns of climate and environmental change. This is reflected in some of the terminology still used. For example the sharp deterioration in climate at the end of the Pleistocene, which began around 12,900 BP and lasted to about 11,400 BP, is known as the Younger Dryas, after the alpine/tundra wild flower *Dryas octopetala* (Mountain Aven) typical of this period.

Other Microfossils

Other microfossils besides pollen and spores may also be used for environmental and climate reconstruction. Important in this respect are diatoms, which are microscopic unicellular algae that live in ponds, lakes, estuaries and the sea. They are sensitive, amongst other things, to water temperature and salinity, and may be used in the reconstruction of regional climate and environmental change.

Fig. 113: Taking a column sample for pollen analysis using monolith tins

Chapter 6: What Remains? – Archaeology

The archaeological remains in Ice Age deposits consist mainly of stone tools. Although Ice Age humans would certainly have made tools from a variety of materials, perhaps especially wood, very few examples of these have survived to the present day. Where they do survive, such artefacts are of exceptional importance.

This chapter introduces the main types of Palaeolithic stone tools found in Ice Age deposits. These range from simple flakes and the cores they were struck from to handaxes, blades, blade cores and pressure-flaked artefacts.

Making Stone Tools

Fig. 114: A hard hammer (quarzite cobble) being used to detach a large flake from a core (flint nodule)

Fig. 115: A soft hammer (antler) being used for final flaking of a handaxe

Stone tools are made by a process known as knapping. Those who make stone tools are called knappers. Knapping involves shaping stones into tools by striking off flakes. The suitable lump or nodule of stone one starts off with is called the core. The pieces removed, be they big or small, are called flakes. The aim may be to reduce, by flaking, the core into a useful tool such as a handaxe. Or the aim may be to produce sharp flakes that can by themselves be used as tools, or be modified by further flaking into tools. The work of modern knappers who replicate ancient technologies using authentic materials and processes has taught us much about early human technologies and adaptations.

During Lower and Middle Palaeolithic times knappers used direct percussion techniques. This means that the core was struck directly with a hammer to detach flakes. Any further shaping of detached flakes was also undertaken using a hammer. Hammers come in different types, with different properties. Knappers will usually select both the hammer and the core carefully, and may switch back and forth between different types of hammer at different stages of their work.

Hard hammers (Fig. 114) are stones, for example quartzite cobbles, that produce relatively thick, large flakes. They are often used in the initial stages of knapping a core. Soft hammers (Fig. 115), made of antler, wood or bone, produce thinner flakes and are generally associated with the final shaping of an artefact.

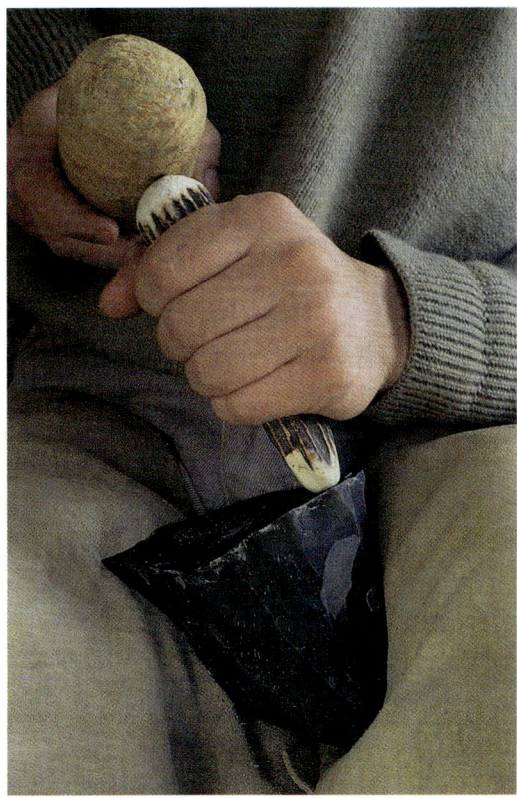

Fig. 116: Indirect percussion using an antler punch

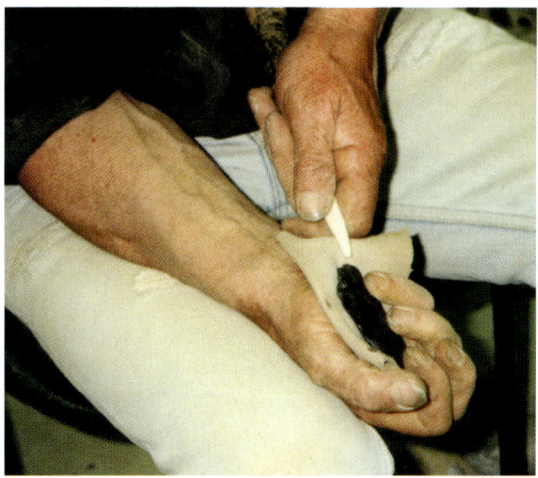

Fig. 117: Pressure flaking

In the Upper Palaeolithic indirect percussion techniques become evident in the archaeological record – here, as the name implies, rather than directly hitting the core to remove flakes, an intermediary object is used (often referred to as punch). This is first seen in the production of Upper Palaeolithic blades, where an antler punch is placed on the exact part of the core where knapper wants to remove a blade. The punch is then struck with a hammer (Fig. 116). The use of punches improves the precision of knapping.

A further Upper Palaeolithic innovation was the technique of pressure flaking (Fig. 117). With this technique a piece of antler or bone is placed on the edge of a detached flake and pressure is applied to detach very small, thin flakes. This technique precisely shaped the flake into the desired end product, for example a Solutrean Leaf Point (Fig. 118).

In Europe, flint was the most widely used lithic (stone) raw material in Palaeolithic times. Flint is a fine-grained, very homogenous stone, similar to glass in its mechanical properties. This means that not only does it create very sharp edges as it fractures, but it does so in a predictable manner. Flint fractures conchoidally (literally 'shell-like'), and as it is knapped a series of recognisable percussion features are produced. These percussion features help to differentiate humanly made artefacts from naturally fractured rocks – even among gravel deposits full of naturally broken stones!

Flint forms in chalk, and both large and small nodules of flint can be found wherever chalk deposits are exposed (Fig. 119). Flint collected from naturally eroding chalk outcrops provided early humans with high-quality raw material for tools. Flint was also collected from secondary sources such as river gravels – although the size and quality of the pebbles available was very variable.

Most, but by no means all, of the Palaeolithic artefacts from England are made from flint. However, flint was not available everywhere in England and Palaeolithic people also used other suitable fine-grained rocks, such as quartzite and Greensand chert, and occasionally less common materials such as andesitic tuff (a volcanic rock).

ICE AGE

Fig. 118: A replica Soultrean Leaf Point finished using the pressure flaking technique

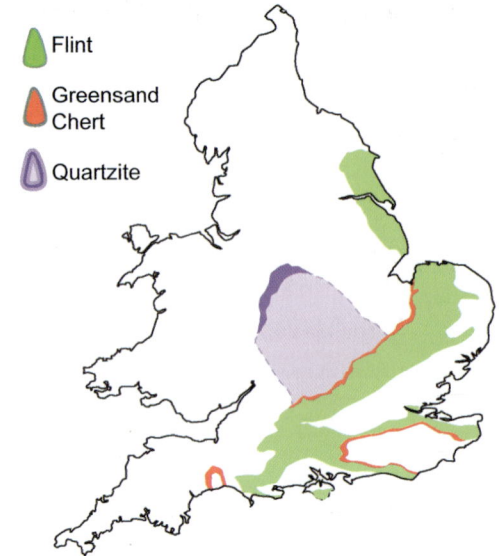

Fig.119: Map showing the distribution of the major sources of raw materials used for making stone tools during the Palaeolithic

The raw materials map (Fig. 119) provides an idea of what materials Palaeolithic artefacts in different parts of the country are likely to be made from, although it is important to remember that non-local 'exotic' materials may also have been utilised.

FLAKES AND CORES

FLAKES

Flakes are the simplest type of lithic artefact. They are quick to produce, very sharp and can easily be retouched into other tool types. Knapping creates a series of recognisable features on both the flake and the core it is detached from. It is these percussion features that allow genuine Palaeolithic artefacts to be recognised.

The example illustrated in Figure 120 is a modern replica of an ancient flake. This means that the flint is fresh and neither stained nor patinated (patination is a mineral coating that forms on the stone over time). Archaeological examples are more likely to be stained from their long exposure to chemicals in the sediments. Orange, brown, yellow and cream are the most common staining colours for flint to develop (Fig. 121). On other raw materials such as chert and quartzite this staining is harder to see due to the natural colouration of the raw material (Fig. 122).

RETOUCHED FLAKES

Freshly knapped flakes (Fig. 123) are extremely sharp, and would have been used for a variety of cutting tasks. However, very thin, very sharp edges quickly become blunted with use. Retouching the edge

What Remains? – Archaeology

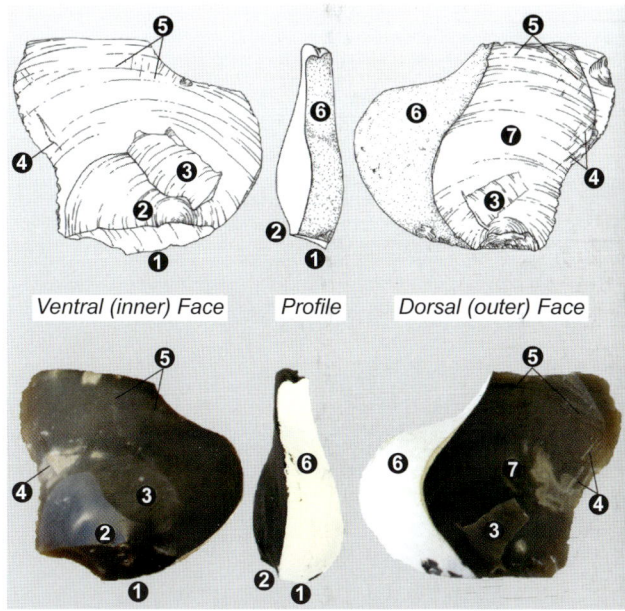

Fig. 120: A flint flake showing percussion features

1 Striking platform: this is the flat area where the hammer strikes the core to remove the flake

2 Bulb of percussion: adjacent to where the hammer strikes, the force of the impact produces a conical swelling

3 Eraillure scar: the scar resulting from a small secondary flake removed as the bulb of percussion formed

4 Radial fissures: small cracks, also known as hackles, pointing to the bulb of percussion

5 Ripples: these radiate from the bulb of percussion, travelling the length of the flake. Ripples indicate the direction of the blow that removed the flake from the core

6 Cortex: the outer 'skin' of a nodule

7 Dorsal flake scars: evidence of previous flaking may be present on the dorsal (outer) surface of the flake, either as ripples or complete negative flake scars which have a hollow where the bulb of percussion of the previous flake formed

Fig. 121: A flint flake stained orangey brown

Fig. 122: A chert flake, naturally dark brown, showing some orange staining

of a flake makes it much more durable and suitable for heavy tasks such as wood working and hide processing, which would quickly blunt a non-retouched flake.

Retouching is the removal of a series of small flakes to modify the shape or edge of an artefact. Retouching may be done by direct percussion using a small hard (stone) or soft (e.g. antler) hammer, or by pressure flaking. All these methods were common in the Upper Palaeolithic but during the Lower and Middle Palaeolithic only direct percussion was used. Figure 124 shows some common types of retouched tool.

Simple Cores

Cores are the pieces of lithic raw material from which flakes are detached (Fig. 125). Just like flakes, cores have a series of diagnostic features that help to distinguish man-made artefacts from naturally fractured stones.

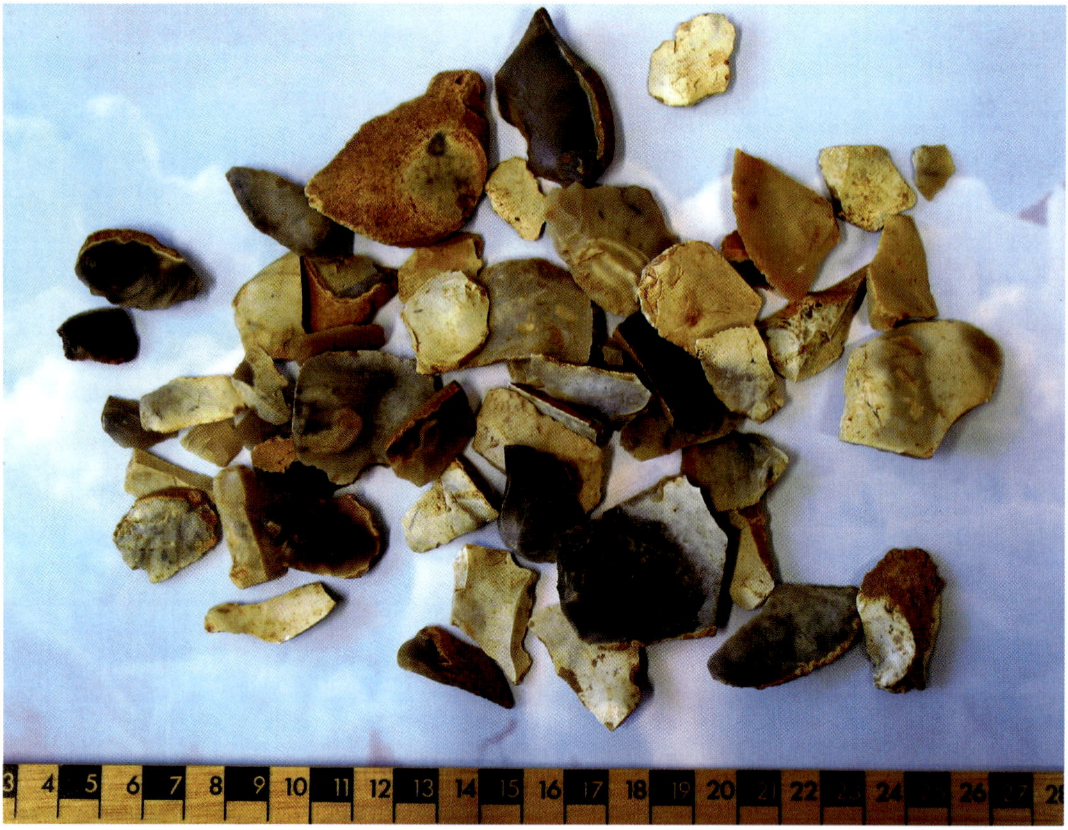

Fig. 123: A variety of flakes produced by knapping

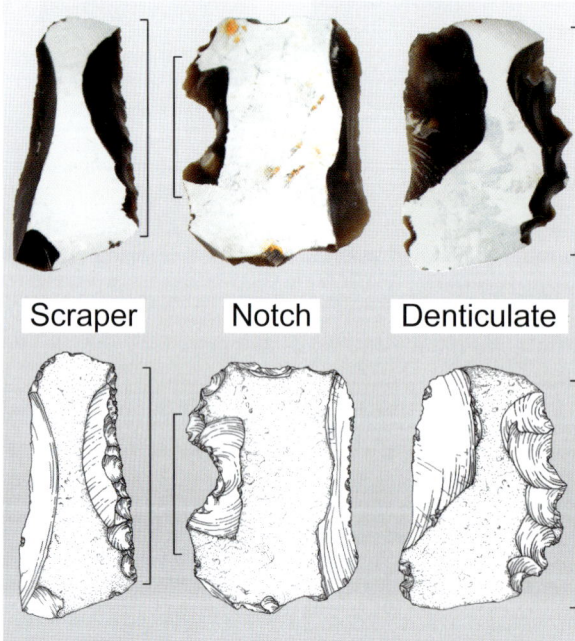

Fig. 124: Some common unifacial retouched tool types

When flakes are detached from a core they leave a distinct hollowed-out impression of their shape in the core. This 'negative' flake scar corresponds to the 'positive' bulb of percussion present on the detached flake, and ripples indicate the direction in which the flake was detached from the core.

Each flake that is removed creates a negative flake scar on the core, and as knapping of the core continues these negative scars may overlap and overprint one another (Fig. 126) – thus the most complete negative flake scars on a core generally relate to the flakes that were removed last.

Débitage, Wear and Refitting

The manufacture of stone tools created a lot of debris as well as the tools that were the purpose of the knapping (Fig. 127). In fact, it usually created much more debris

What Remains? – Archaeology

Fig. 125: A sequence of flakes being removed from a flint core

SAMPLE CORE

Flake 1 was removed first. The core was then rotated 90° and the scar of the first flake used as a platform for flake removals 2 and 3. The core was then re-rotated 90° and flakes 4 and 5 removed from the core. Further flake removals from the same location as 4 and 5 would have been possible.

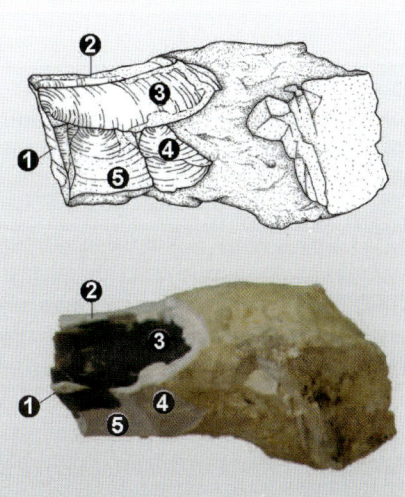

Fig. 126: A flint core showing the percussion features arising from the removal of a sequence of flakes

Fig. 127: A prepared flake (left) and a waste flake (right)

than it did tools. This debris is often known as débitage.[1] The tiny flakes resulting from retouching are known as micro-débitage.

Débitage is as much evidence of past human activity as are the tools themselves. It bears evidence of human manufacture (e.g. flake scars, bulbs of percussion) in the same way as the tools. Indeed, while it may be possible to determine that material has been humanly struck, it is often not possible to determine which pieces were intended as tools and which were not (if such a clear distinction existed in the mind of the knapper).

Close inspection of pieces may reveal whether they were actually used as tools or not, by showing damage and wear, and in some cases it is possible to suggest the use to which a tool was put. Different uses, for example working wood or cutting meat, leave different patterns of wear on the edges of a tool. This can be best seen under a microscope and is known as microwear. By carrying out experiments using replica tools for different tasks and then studying the patterns of wear that arise, it is sometimes possible to suggest the use to which an ancient tool was put. This is generally only possible in those rare instances where tools have been little disturbed since the time of their discard. Such tools are described as surviving in a primary context. For microwear to survive it is also necessary that patination has been minimal.

Fig. 128: Refitting: a large flint nodule (length c. 30cm) rebuilt from flakes found at Boxgrove. The cavity probably represents a handaxe roughout.

Usually tools have been disturbed and transported by natural forces such as the flow of rivers. In these cases the tools will have been bashed by other stones, and their edges blunted and smoothed so that any former trace of microwear has been removed. Tools that show clear evidence of such bashing and smoothing due to transport in a river are described as 'rolled'. They have survived only in a secondary context.

In rare cases at primary context sites, the debris from stone tool making may survive more or less where it fell. It is then possible to reassemble the flakes struck from a core into the original shape of the core. This process is known as refitting (Fig. 128). Although refitting is difficult and time consuming, it enables archaeologists to reconstruct precisely the actions of the knapper. This provides detailed information on how the tools were made, and also provides some insights into the minds of the knappers.

---

[1] Strictly speaking, this usage is incorrect. The French term *débitage* means the process of knapping a core to produce usable pieces.

The Modes of Stone Tool Manufacture

As has already been seen, many different techniques may be used in stone knapping and many different types of tool can be made. One way of bringing order into this diversity is to describe the basic 'mode' of stone tool manufacture employed.

The simplest mode of manufacture, called Mode 1, is the striking of flakes from a core. The flakes may be used just as they are or modified by further flaking or retouch for particular purposes. The same is true of the core.

If a global perspective is taken of the development of stone tool technology, Mode 1 is the earliest technology to appear; tools made by this technique date back to around 2.6 million years ago in Africa. Later, over the course of the Pleistocene, more complex modes of stone tool production appear in sequence, called Modes 2, 3 and 4. These will be described below.

However, an important point is that the mode of stone tool technology employed is not always an indicator of date. Mode 1 technology, with its simple flakes and flake pebble tools, continued to be employed after the introduction of Mode 2 technology (handaxes), and may be found throughout prehistory. Likewise Mode 2 technology continued to be used after the introduction of Mode 3 ('prepared-core') technology, and so on.

Handaxes (Mode 2)

The most common type of Palaeolithic tool recovered in this country is the handaxe. The oldest handaxes known from this country are over 500,000 years old. Handaxes, albeit of a characteristic small type, were still being made by Neanderthals around 50,000 years ago.

How Were They Made?

Handaxes are also known as bifaces, and this alternative name refers to how they were made – these tools were extensively shaped and flaked on both faces to create a long and tough cutting edge (Fig. 129). The production of handaxes and similar tools displaying extensive shaping defines Mode 2 technology.

Handaxes dominate the Lower Palaeolithic archaeological record because they are large, durable and due to their extensive working they are easily recognisable as humanly-made artefacts (Fig. 130). As a result they have been much collected over the years by antiquarians, archaeologists and quarry workers.

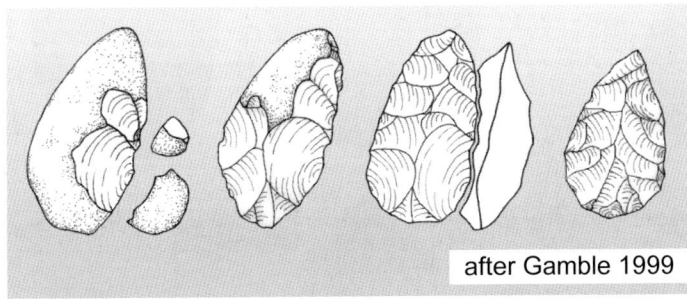

after Gamble 1999

Fig. 129: Core reduction method of handaxe production

What Were They Made From?

The handaxes that have been recovered in this country were made from a range of raw materials including andesitic tuff, quartz, quartzite, chert and flint (Fig. 131). The raw materials map (Fig. 119 above) provides an indication of the dominant raw material available in different areas of the

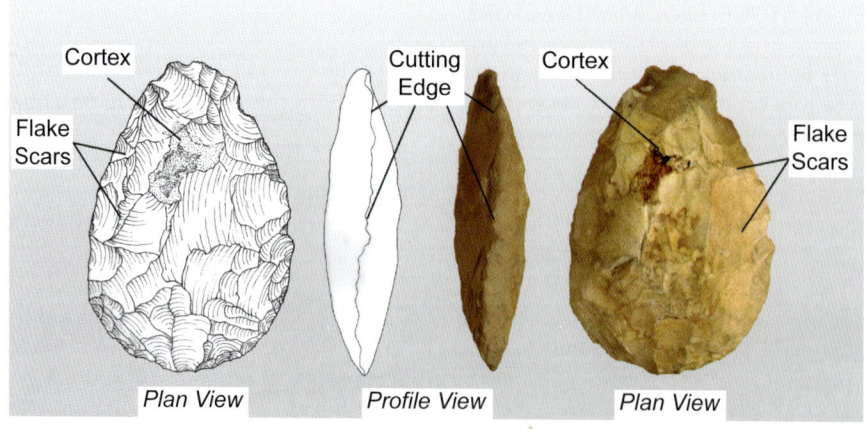

Fig. 130: A lightly stained flint handaxe showing some of the principal features

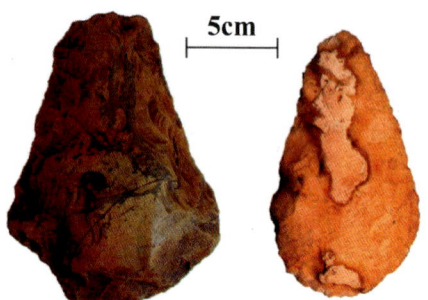

Flint handaxes occur in many parts of England. Commonly they are iron-stained, as shown above. Handaxes from gravels may also show abrasion and/or damage similar to the handaxe on the left.

Quartzite cobbles suitable for handaxe manufacture are scattered throughout central England. Handaxes are commonly made on split quartzite cobbles like those shown above.

Greensand Chert is mechanically similar to flint, though it often looks much coarser. The largest numbers of Greensand Chert artefacts occur in the southwest of England.

Andesitic tuff is an ancient volcanic deposit suitable for stone-tool manufacture. Handaxes of andesitic tuff have been found mainly in the Midlands.

Fig. 131: Examples of handaxes of different materials found in gravel deposits around England

WHAT REMAINS? – ARCHAEOLOGY

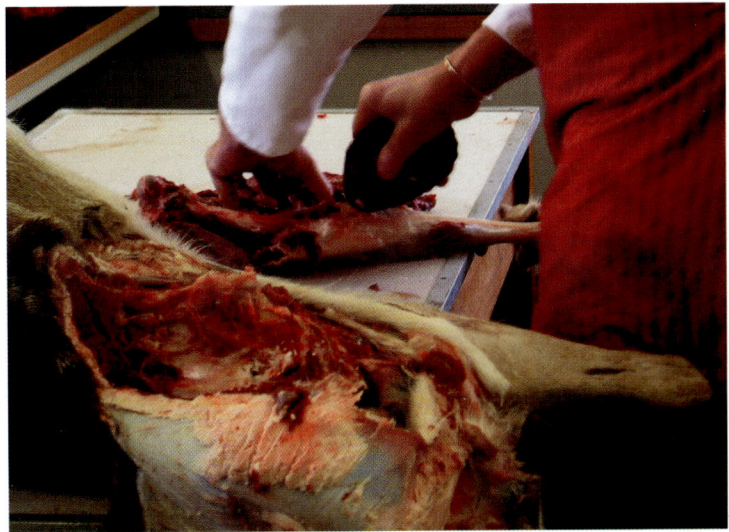

Fig. 132: Handaxe being used experimentally for butchery and skinning

country. However, in smaller quantities knappers also used 'exotic' stone types that were not so readily available in their local area, especially if these possessed the flaking qualities they were looking for.

WHAT WERE THEY USED FOR?

Handaxes are considered to be multipurpose tools. Many experiments have shown them to be highly efficient butchery tools (Fig. 132). They can be easily resharpened, which both extends the useful life of the handaxe and can also provide a source of sharp flakes if required.

HANDAXE SHAPES AND SIZES

Handaxes come in a range of sizes. Occasionally very large examples (over 30cm long) and much smaller specimens (approx. 5cm long) are found, although as the name implies most are of a size that fits comfortably into the hand. There is no evidence that handaxes were ever hafted (i.e. fitted with a handle).

Handaxes are also found in a range of shapes and exhibit differing degrees of 'refinement'. It was once thought that handaxe forms evolved through time and that crudely shaped examples made with just a hard hammer were older than more finely-flaked forms, finished with a soft hammer. However, sites such as Boxgrove in West Sussex with 'finely' made handaxes of an early date (about 500,000 years old), finally disproved this theory. It is now recognised that 'refinement' is not a useful indicator of the age of a handaxe (outside of Africa at least).

Variations in handaxe shape and refinement are now generally considered to occur for other reasons. These might include the quality of the raw material and the size of the nodules or cobbles available, the function of the tool, the skill of the knapper, and perhaps short-term local traditions. Some of the most common handaxe shapes, such as *pointed, ovate* and *chordate* (heart-shaped) are shown in Figure 133. Palaeolithic handaxes are not to be confused with Neolithic stone axes (also found in quarries), which are much later in date, different in shape, and often ground or polished.

CRUDE   POINTED   OVATE   SUB-CORDATE   CORDATE   CLEAVER   FICRON   BOUT COUPÉ

Fig. 133: Some of the handaxe shapes that have been defined

Assemblages and Industries

Two important concepts in the study of stone tools are those of assemblages and industries. An assemblage is a collection of artefacts (or fossils) found together at a particular site and thought to be similar in date (for example they may all come from the same soil layer). A stone-tool assemblage, sometimes containing thousands of items, will usually contain tools of different types and often the débitage from their manufacture also. The tools may have been made using different modes of technology.

Where similar assemblages of stone tools are found at a wide number of sites, together they are taken to define a stone-tool 'industry'. Industries are usually named after the sites where they were first recognised and defined. Thus Lower Palaeolithic assemblages containing handaxes are called Acheulean, after a quarry near St. Acheul in northern France. The Acheulean is a very widespread industry, found throughout much of the Old World except east Asia. In Britain another Lower Palaeolithic industry, lacking handaxes and comprising only pebble core and flake tools (Mode 1), is called the Clactonian, after the site of Clacton-on-Sea in Essex.

The Levallois Technique (Mode 3)

Named after the Parisian suburb where it was first identified, the Levallois technique refers to a specific type of 'prepared-core' technology. The appearance of this technology marks the transition from the Lower to the Middle Palaeolithic and in Europe it is mainly associated with the Neanderthals.

Prepared core (Mode 3) technologies are more 'sophisticated' than the other technologies we have described. They involve carefully preparing a core from which a flake or flakes of pre-determined size or shape will then be removed (Fig. 134).

In the Levallois technique, a striking platform is created at one end of the core and preparation flakes are removed from all around the edge of the core. In this way the shape of the Levallois flake to be removed is defined. This process creates a domed shape to the upper half of the core, with a distinctive pattern of scars similar to the shape of a tortoise shell. Levallois cores thus are often called tortoise cores (Fig. 135).

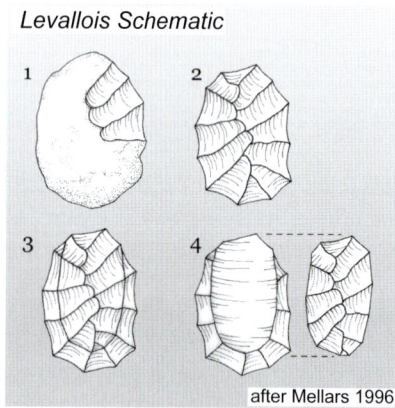

Levallois Schematic

after Mellars 1996

Fig. 134: A schematic diagram showing the preparation of a Levallois 'tortoise' core and detachment of a Levallois flake

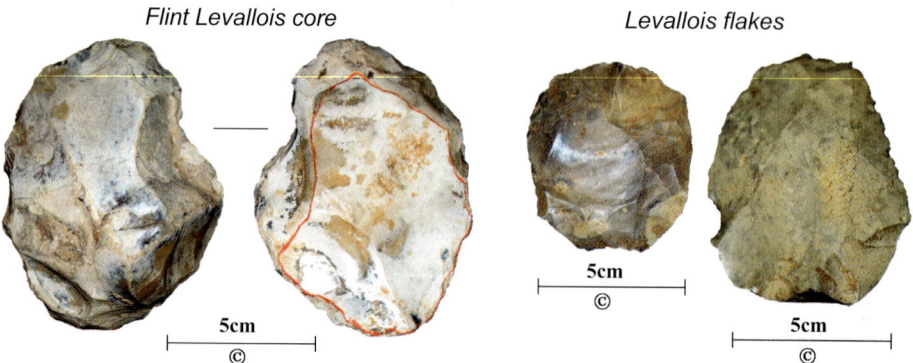

Flint Levallois core — Location of *Levallois flake* removal marked in red

Levallois flakes

Fig. 135: A Levallois 'tortoise' core and Levallois flakes

WHAT REMAINS? – ARCHAEOLOGY

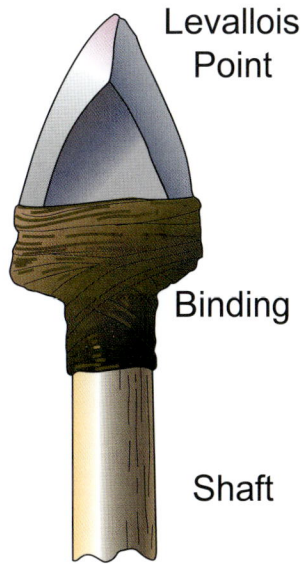

Fig.136: Reconstruction of a Levallois point hafted to form a spear tip.

Once all the preparation flakes have been removed, a final removal creates the Levallois flake, the shape and dimensions of which have been determined by the preparation of the core. Levallois cores can be prepared in such a way that the final flake is triangular – these are known as Levallois points. Levallois cores can also be prepared so that several successive Levallois flakes can be removed.

Levallois flakes (and points) have a distinctive plano-convex profile. That is, they are flat on their lower (ventral) face, where the flake has been removed from the core, and convex on their upper (dorsal) face. The preparation of the core also creates an easily recognisable dorsal scar pattern.

Levallois flakes and points were retouched into a wide range of tool forms. Levallois points may have been hafted, using a resin glue and sinew binding, onto a wooden haft to make stone-tipped spears (Fig. 136). The hafting of stone tools may have been one of the most significant developments of Mode 2 technology. Earlier wooden spears (e.g. the example from Clacton, Fig. 29, p.33) were made just by sharpening the point, although this may have been fire hardened.

THE MOUSTERIAN INDUSTRY

The Mousterian is the most widespread Middle Palaeolithic stone-tool industry found in Europe. It is named after Le Moustier rockshelter in south-west France, where excavations began in 1863. Mous-

Fig. 137: Mousterian tools – point and side-scraper

terian artefacts were studied and classified in great detail by the renowned French Palaeolithic archaeologist François Bordes. Assemblages of Mousterian artefacts sometimes (but not always) include tools made by the Levallois technique, but they also include tools made by other methods. Bordes recognised more than sixty different tool types, just two of which are illustrated here (Fig. 137).

In any particular Mousterian assemblage different types of tools are present or absent, and those present occur in different proportions. From the study of these differences Bordes defined five 'Mousterian variants'. What these variants mean has been the subject of much debate. Bordes himself considered them to represent five different tribes. Others have believed them to represent the debris from carrying out different kinds of activities, for example at different times of the year, which would require the use of different types of tools in varying proportions. There is also a chronological dimension, with some variants being found consistently later than others at a number of different sites.

In England, most of the known Late Middle Palaeolithic sites (after around 60,000 BP) can be classified as belonging to a Mousterian variant called the Mousterian of Acheulian Tradition (MAT). This means that the assemblages, which display little use of the Levallois technique, include handaxes. These handaxes are of a very specific flat-based, chordate (heart-shaped) type known as *bout coupé* (see Fig. 133 above). They are one of the most characteristic markers of a Neanderthal presence in Britain.

Upper Palaeolithic Technologies (Mode 4)

Upper Palaeolithic technologies are associated with the colonisation of Europe by modern humans (see chapter 3) and in this continent span the period from approximately 45,000 BP to around 11,400 BP. The most important innovation found in Upper Palaeolithic lithic technology is a shift to systematic blade production (Mode 4). Blades are defined as flakes that are more than twice as long as they are wide, and although some blades were made in earlier periods it is only in the Upper Palaeolithic that blade technology comes to dominance.

Upper Palaeolithic blades are thin, narrow and long. To produce them the core needs to be carefully prepared by flaking to create a crested 'keel' down

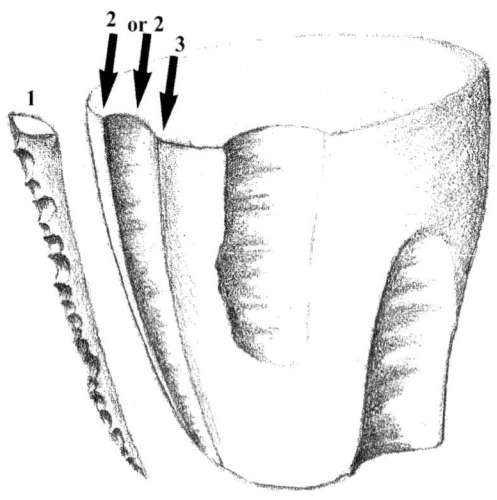
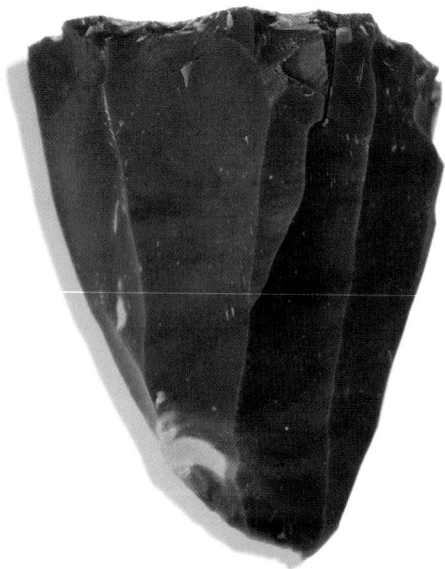

Fig. 138: Preparation of a blade core and detachment of the crested blade, setting up the core for detachment of further blades

Fig. 139: Blade core showing removal facets

WHAT REMAINS? – ARCHAEOLOGY

Fig. 140: Blades struck from a single core

one of the long sides of the core. A striking platform is made at one end (or sometimes both ends) of the core, and the crested blade is detached (Fig. 138). This creates a pair of parallel ridges and these are used as the starting point for the removal of further blades, working round the core (Fig. 139). A soft hammer or a punch (giving greater control) may be used to strike off the blades (Fig. 140; Fig. 116, p.89).

Blade production maximises the amount of useful tool edge that it is possible to extract from a block of raw material, making it a highly efficient technology. Also, the blades produced are of more standardised dimensions than the flakes by other technologies.

Blades were retouched into an unprecedented number of different types of tool, ranging from simple scrapers to projectile points, borers, piercers and leaf points (Fig. 141). Blades could be mounted in handles or on spear shafts, and when they broke they could easily be replaced.

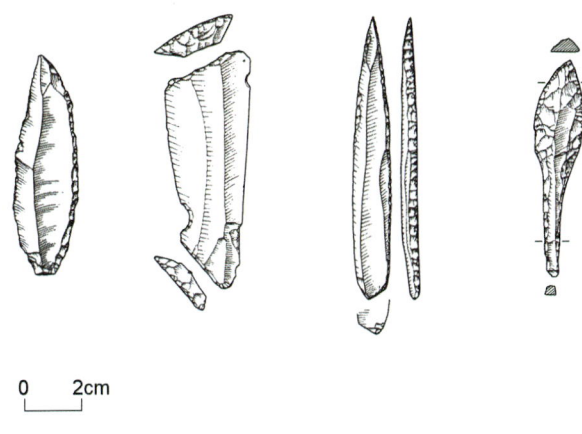

0   2cm

Fig. 141: A range of earlier Upper Palaeolithic tools mainly made from elongated blades, and comprising missile heads and an insert for a hafted knife

The different types and styles of tools made on blades show clear regional and chronological variation, and change is now much more rapid than in earlier periods. Archaeologists have recognised a series of major industries that follow each other in succession, the most well known being named after sites in France. These include the Aurignacian, Gravettian and Solutrean in the Early Upper Palaeolithic (up to the Last Glacial Maximum, when the ice sheets were at their greatest extent) and the Magdalenian in the Late Upper Palaeolithic. Tools belonging to variants of these industries are found in Britain, for example the Creswellian (named after Creswell Crags) is a variant of the Magdalenian.

NON-LITHIC ARTEFACTS OF THE UPPER PALAEOLITHIC

A common type of Upper Palaeolithic tool made on flint blades is the 'burin'. These are chisel-ended tools particularly associated with the manufacture and carving of the wide range of bone and antler artefacts that are one aspect of the cultural and technical diversity of the Upper Palaeolithic (Fig. 142).

Occasional bone and wood artefacts are known from Lower and Middle Palaeolithic sites. However, it is only in the Upper Palaeolithic that the use of bone, ivory and antler becomes widespread.

Bone points, needles, spear throwers and harpoons (Fig. 143) testify to the improved technologies of Upper Palaeolithic people. Beads and carved figurines are an indication of their much more complex social and cultural lives compared to their predecessors. Such 'portable' art is rare in Britain but include an engraving of a horse on a rib fragment from Robin Hood Cave, Creswell Crags (Fig. 144).

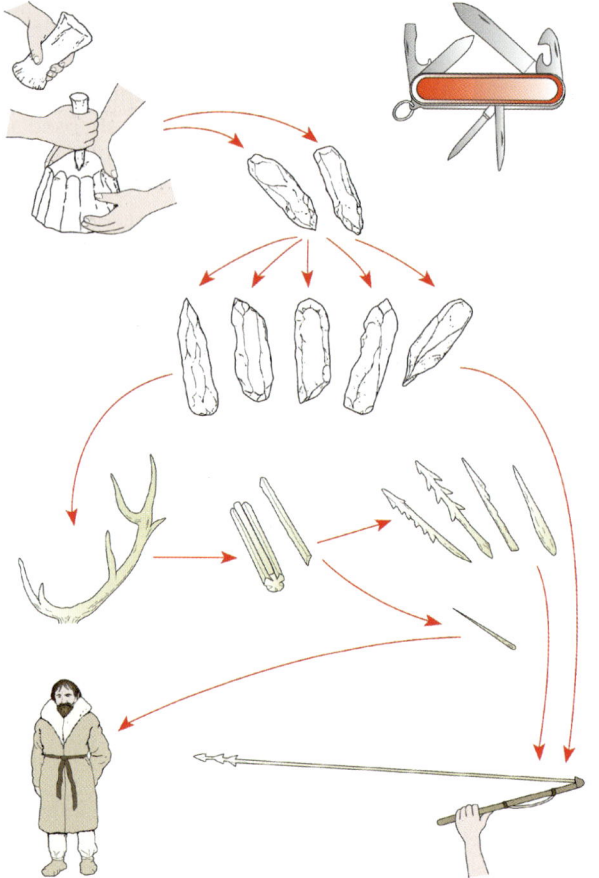

Fig. 142: The uses of Upper Palaeolithic technology: from core, to blade, to blade tool, to antler tool. The versatility of this technology has been compared to a Swiss Army Knife.

Fig. 143: Upper Palaeolithic antler harpoon

Fig. 144: A rib fragment from Robin Hood Cave, Creswell Crags engraved with the image of a horse's head

IDENTIFYING STONE TOOLS – A CHECKLIST

This chapter has described the main types of Lower, Middle and Upper Palaeolithic artefacts that may be encountered in a sand and gravel quarry. As we have seen, stone tool technology changes dramatically through time. However, there are some general principles that can be used to successfully distinguish humanly-made stone artefacts of any period from naturally fractured stone.

Bulb of percussion: this is the swelling on a flake (or blade) that occurs just below the place where the core was struck to remove the flake. It is easiest to see on flint flakes but also occurs on flakes of other raw materials, such as quartzite.

Flake scars: the evidence on an artefact for earlier flake removals. They are made up of ripples and a 'hollowed-out' negative bulb of percussion. Multiple negative flake scars help to distinguish between deliberate knapping and natural processes (e.g. rocks colliding and fracturing in a river), which do not usually create a series of complete negative flake scars.

Retouched flakes: the presence of small flake scars (retouch) that alter the shape and/or the angle of the edge can be a good indicator of genuine artefact status. However, when dealing with Ice Age river gravels it is always worth remembering that small flakes may also have been removed by natural processes.

Handaxes: handaxes or bifaces are among the easiest types of artefact to identify, due to the large amount of shaping and flaking involved in their manufacture.

Artefact size and shape: Palaeolithic artefacts come in a wide range of sizes and shapes. As a very rough guide, handaxes are generally approximately hand-sized, and flakes somewhat smaller. Cores can be any size. Due to the ferocity of Ice Age rivers it can be very difficult to prove that very small pieces are artefacts as the features they display may well result from impacts with cobbles.

Artefact colour and physical condition: many artefacts from Ice Age rivers will show iron staining, of an orange-type colour, or patination. Artefacts may also have been damaged by river transportation; most typically this shows as abrasion to the ridges between the negative flake scars and/or chipping to the edges of the piece.

Chapter 7: How Old Is It? – Dating the Ice Age

A very wide range of scientific techniques can now be used in the attempt to accurately date Ice Age deposits. Here we will describe only those techniques that are usually used to date the types of Pleistocene deposits that typically occur in British sand and gravel quarries (brief details of other techniques will be found in the glossary).

There are two broad categories of dating – relative dating and numerical dating (also known as absolute dating). Relative dating involves getting things – geological deposits, assemblages of fossil remains and stone tools – in the right sequence. The first requirement is to get the deposits in the right sequence at any given site or locality. The second requirement is to determine the relative dates of deposits and assemblages at different localities. Are two deposits at different localities of the same date, or is one earlier or later than the other?

While relative dating is about getting things in the right order, it does not of itself provide information about how old things are in terms of calendar years. This is the job of numerical dating techniques, which provide an estimate of the actual age of a deposit or assemblage, usually, for the Pleistocene, expressed in years before present (BP). For radiocarbon dates 'the present' is actually 1950, the time when the radiocarbon technique was first introduced. Obtaining numerical dates for deposits is of crucial importance because it is only in this way that deposits can be related to the 'master framework' provided by the oxygen isotope records from deep-sea and, more recently, ice cores.

In practice, relative and numerical dating techniques are used together. For example, if a numerical date is obtained for a particular deposit – say a gravel terrace or a characteristic animal bone assemblage – at one locality, then if the same deposit or a similar animal bone assemblage is found at another locality, it may be inferred (with caution) that they are of the same date. This is known as 'cross-dating'.

Recent advances in both relative and numerical dating techniques have transformed our understanding of Ice Age Britain, particularly through relating geological deposits, river terraces and assemblages of animal bones and stone tools to their appropriate marine isotope stage. On the whole, many deposits, assemblages and events have turned out to be earlier – sometimes hundreds of thousands of years earlier – than was previously thought. This means that many of the dates given in older books, and also in some recent ones, are now believed to be wrong.

Despite these advances, much work remains to be done to refine existing dating techniques and develop new ones, so that more reliable and accurate dates can be obtained. This is a very active area of research. Most dating methods involve margins of error intrinsic to the technique as well as the possibility of more serious errors occurring as the result flawed sampling or analysis. Confidence is considerably increased when two or more techniques, based on different principles, converge on the same date.

Relative Dating

The fundamental principle behind most relative dating techniques is stratigraphy. This is the geological concept that deposits are laid down in a sequence, usually with the oldest deposits at the bottom and the progressively younger ones stacked above. This does not always hold true, however, as younger sediments can intrude into older ones from both above and below. The most important example of this in the present context is the action of rivers, which progressively deepen their valleys, cutting into older deposits and laying down younger deposits below the level of the older ones. It is this processes that causes the phenomenon of terrace 'staircases' up the sides of river valleys, the terraces becoming progressively younger as one moves down the side of the valley (see chapter 2). *Within* each terrace, however, the deposits will have been laid down in the normal sequence – oldest first and lowest down.

Furthermore, natural processes that occur subsequent to the laying down of deposits in orderly stratigraphic sequences can distort the true sequence. During the Ice Age, the principal agent of such distortion is glaciers, which can pick up whole chunks of landscape, transport them somewhere else and even turn them upside down – a process known as 'rafting'. Interpreting Ice Age stratigraphic sequences is therefore by no means straightforward and is the specialist task of geomorphologists (scientists who study land forms).

One important aspect of relative dating of Ice Age deposits is the recognition of deposits laid down by particular glaciations. This is not always easy because the deposits relating to early glaciations can be very patchy, much of the evidence being destroyed by the action of later glaciers. The composition of the till (or boulder clay – see chapter 4) deposited by ice sheets can however provide a 'signature' that allows scattered, patchy deposits to be related to the same event. Only very recently an early ice advance, previously unknown, has been recognised through evidence of this nature. It is termed the Happisburgh Glaciation and is thought to date to around 650,000 BP. Not all are convinced by the evidence for the Happisburgh Glaciation, however, although the question of its existence and the stratigraphic relationship of deposits attributed to it containing very early stone tools is crucial to the understanding of the earliest human occupation of Britain.

BIOSTRATIGRAPHY

Biostratigraphy is a relative dating technique of great importance to the understanding of Ice Age Britain. It is based on the principle that the composition of the fauna of the British Isles has changed through time, and that the sequence of different faunas can be used as a relative dating technique.

The primary agent causing the fauna of Britain to change through time has of course been climate change, and in particular the sequence of glacial-interglacial cycles the climate has undergone. During glacial periods warmth-loving species retreated to 'refugia' in southern Europe, recolonising Britain when warm interglacial conditions returned. However, it was not just a simple pattern of the ebb and flow of the same warm- and cold-adapted species over hundreds of thousands of years. Amongst the mammals, evolution (and its necessary counterpart, extinction) was underway, also driven in large part by climate change. So the composition of the British fauna in each different cold or warm stage was also different, sometimes subtly, sometimes not so subtly.

Furthermore, large-scale patterns of species' evolution and extinction were not the only factors at play. There were also more localised, historically specific factors. For example, some interglacials were warmer than others. The Pakefield (around 700,000 BP) and the Ipswichian (around 125,000 BP) were particularly warm interglacials and in these alone the hippopotamus (albeit different species) was a prominent member of the British fauna; in other interglacials it was absent. Major glaciations reshaped the landscape and the courses of rivers, meaning that the landscape and hence the ecology of Britain were never quite the same between two broadly similar climatic periods. The question of whether Britain was or was not an island during a particular interglacial will have had profound implications for the pattern of occupation by land animals (including humans).

The combined effect of all these factors (and others) was to give the fauna of each climatic episode in Britain's long Ice Age history a distinctive character. To turn this into a relative dating technique is basically a matter of defining the fossil fauna characteristic of each period, based on the presence or absence of particular species, the evolutionary stage of particular lineages and a number of other factors. Then when new deposits are found, for example in a gravel quarry, containing a good assemblage of faunal remains, these deposits can be slotted into the framework established. Furthermore, it is possible to use the technique to provide new dates for faunal assemblages (and associated stone tools) excavated in the past, before the biostratigraphical technique had become properly developed.

The full potential of biostratigraphy as a dating technique is only realised when it is used in combination with numerical dating techniques as well as geological stratigraphy, including river terrace sequences. Numerical dating enables distinctive faunal assemblages to be related to specific marine oxygen isotope stages and sometimes shorter climatic periods within them.

To formalise the biostratigraphic approach scientists have defined what are called Mammal Assemblage Zones (MAZs). Each MAZ is named after a type locality where the assemblage is defined, for example the Banwell Bone Cave MAZ (MIS 4) or the Gough's Cave MAZ (late MIS 2).

While biostratigraphy is based on the principle of comparing and arranging in relative order different whole assemblages of animal fossils, certain species are of particular value because the dates at which they evolved or went extinct are broadly known. Particularly important amongst these 'indicator species' are certain species of voles. For example, between around 600,000 BP and 500,000 BP an extinct vole called *Mimomys savini* was replaced by its descendant, the water vole *Arvicola terrestris cantiana*. The molar teeth of the two species are distinctive, the former with closed roots and the latter open rooted.

Some of the earliest sites in Europe with evidence of human occupation, such as Gran Dolina in Spain (around 800,000 BP) and Pakefield (around 700,000 BP), are associated with *Mimomys* while later sites such as Boxgrove (around 500,000 BP) are associated with *Arvicola*. This method of dating is often referred to as the Vole Clock. However, biostratigraphical methods of dating may not always be in sync with other dating methods, and local patterns of the appearance and disappearance of particular species may complicate the large-scale patterns. As with all dating methods relating to the Ice Age, more research and refinement is always needed.

Numerical Dating

Most scientific numerical dating methods rely on making use of various kinds of chemical change that naturally occur at known rates in various substances. Something, for example the death of an organism, triggers the process of chemical change to begin. If the rate of chemical change is known and the amount of chemical change can be measured, then an estimate of the date at which the process was triggered can be obtained.

Several assumptions are involved in each technique. For example, problems can occur if the rate of chemical change is in fact variable, and measurement of the amount of change that has taken place can never be entirely accurate. For reasons such as these the dates obtained are statistical probabilities covering a span of time (which may be thousands of years long) rather than precise dates.

Usually more than one date is obtained and statistical methods are used to refine the probability range of the dates. Particularly useful are circumstances where dates are obtained from different levels in a stratigraphic sequence or where different techniques are used in combination. These provide 'internal' and 'independent' checks on the consistency of the dates.

Below we outline the three techniques – radiocarbon dating, OSL and AAR – most applicable to dating Ice Age deposits in sand and gravel quarries. However, scientific dating is a fast-moving field, and it is to be expected that other techniques will be applied in the future. One new technique which holds considerable promise, especially for dating the early parts of the British Lower Palaeolithic, is 'cosmogenic radionuclide dating' (a brief explanation is in the glossary).

Radiocarbon Dating

Radiocarbon dating is the best known of a number of dating techniques that use as their 'clock' the rate of decay of unstable radioactive isotopes (isotopes are versions of elements with different atomic weights). Radioactive elements decay at an exponential rate. The time taken for half of the original number of atoms of a radioactive isotope to decay is called its half-life; after two half-lives, one quarter of the atoms will remain, after three half-lives one eighth, and so on until the number of atoms remaining is too small to measure accurately.

The half-lives of different radioactive isotopes are very different. For some the half life is a fraction of a second, for others it is billions of years. The date range that can be covered by a technique that relies on the decay of a radioactive isotope is in large measure governed by the length of the half-life of that isotope.

Carbon-14, the radioactive isotope used in radiocarbon dating, has a half-life of 5,730 years. This makes it very useful for archaeologists studying the last ten thousand years or so, but the technique cannot be used to date material much beyond about 50,000 years ago because in material older than that date too little carbon-14 will remain to measure accurately. Thus the radiocarbon technique can only be used to date the last part of the Ice Age, and most of the Ice Age lies beyond its reach.

Carbon-14 is formed when cosmic radiation reacts with nitrogen atoms in the atmosphere. It is absorbed by plants through carbon dioxide and by animals through feeding off plants or other animals. When the plant or animal dies, carbon-14 uptake ceases and the unstable isotope begins to decay at the rate of 50 per cent every 5,730 years. The amount of carbon-14 remaining in an ancient organic sample (charcoal, bone, etc.) is measured in the laboratory and an estimate of the age of the sample is obtained. In other words, the death of an organism triggers a process of chemical change (the decay of carbon-14) which occurs at a known rate, and the amount of change is measured to obtain an estimate of the date when the process was triggered.

When the radiocarbon technique was first introduced quite large samples were needed to measure the amount of carbon-14 remaining (10-20g of wood or charcoal, or 100-200g of bone). In the 1980s a new technique for measuring the amount of carbon-14 in a sample, called accelerator mass spectrometry (AMS), was introduced. This technique allows much smaller samples of just a few milligrams to be analysed. A more recent refinement (from late 2000) has been ultrafiltration of samples from archaeological bone. Ultrafiltration removes from the sample molecular contaminants which may distort the results. This is particularly important when dating bones from the Late Middle Palaeolithic and Early Upper Palaeolithic, and the recent application of the technique has led to serious revision of the chronology for these periods.

When the radiocarbon technique was developed in the 1950s it was assumed that the concentration of carbon-14 in the atmosphere had remained constant through time. This was an important assumption because if the concentration had been different in the past, age estimates based on the amount of carbon-14 remaining in a sample today would be wrong. For example, if organisms were exposed to higher levels of atmospheric carbon-14 in the past, the carbon-14 content of samples measured today would be higher than expected on the assumption of a constant atmospheric concentration, and misleadingly recent or 'young' age estimates would be obtained.

This is indeed what happened. It was appreciated in the 1960s that the atmospheric levels of carbon-14 had fluctuated in the past. When radiocarbon dates were obtained from tree-rings, the precise calendar date of which can be determined by counting the rings, it emerged that from before about 3000 BP the radiocarbon dates were increasingly too young. Furthermore, the atmospheric concentration of carbon-14 had evidently fluctuated in rather complex ways, probably due to changes in the Earth's magnetic field. Fortunately, tree-ring dating (dendrochronology) also provided a means to systematically correct or 'calibrate' radiocarbon dates. In the scientific literature, calibrated radiocarbon dates are distinguished from uncalibrated ones by having the abbreviation 'cal' written before the 'BP'.

Unfortunately, tree-ring calibration of radiocarbon dates does not extend back into the Pleistocene. However, high-precision uranium-series dates (another dating technique based on the decay of radioactive elements but not generally applicable in the context of sand and gravel quarries) obtained from cores into ancient coral reefs have enabled the calibration of radiocarbon dates to be taken back to around 40,000 BP, close to the limit of the radiocarbon technique. It has been shown that uncalibrated radiocarbon dates between 18,000 BP and 40,000 BP are in general about 3000 years too young. Unfortunately, as yet there is no agreement on the calibration of ages greater than about 11,000 BP. The calibrated dates given in this book relating to the Upper Palaeolithic should therefore be considered only to be indicative.

Optically Stimulated Luminescence Dating (OSL)

Whereas radiocarbon dating is used to date organic materials such as charcoal and bone, OSL is used to date quartz (sand) grains. This makes it potentially an ideal technique for use in quarries, where it can be used to date when stratified deposits of sands and gravels were laid down.

The technique makes use of the fact that all sediments contain tiny amounts of natural radioactive isotopes that emit radiation. Quartz grains have defects in their crystal structure that capture electrons disrupted from their stable atomic position by this radiation. Over time, increasing numbers of electrons accumulate in these 'electron traps'. However, just a few seconds or minutes of exposure to sunlight, such as will happen when the sand is transported by wind or water, serves to empty the electron traps. This sets the clock back to zero. Then, when the sand is laid down and becomes buried, the electrons start to accumulate in their traps once again.

The total number of electrons that have accumulated in the grains since they were buried should be in proportion to the total radiation dose received since the clock was set to zero. This provides a means of estimating the time that has elapsed since the grains were buried.

Samples of sand are collected in the field using light-proof tubes. A gamma spectrometer is then inserted into the sample cavity to measure the amount of natural radiation present (Fig. 145). In the laboratory (Fig. 146), measurement of the number of electrons that have accumulated in the sample is made indirectly through exposure of the isolated quartz grains to controlled lighting and measuring the light (luminescence) emitted, a phenomenon resulting from the recombination of the trapped electrons into their stable atomic position. This process is known as optically stimulated luminescence (OSL).

In order to calculate the age of the sample the total dose of radiation is divided by the estimated annual dose of radiation occurring in the deposit.

A principal source of inaccuracy in optical dating derives from the possibility that the water-transported

How Old Is It? – Dating the Ice Age

Fig. 145: A gamma spectrometer being used in the field to measure the natural radiation present in a sediment

Fig. 146: Analysing OSL samples in the laboratory

quartz grains may not have received sufficient exposure to sunlight to completely empty the 'electron traps' in the crystal lattice. This can result in overestimates of age. However, techniques have recently been developed to identify such partially 'bleached' grains, which may then be eliminated from the analysis. Furthermore, the samples are usually collected in a stratigraphic sequence and should thus produce a chronological sequence of dates; this also helps the identification of 'rogue' dates.

AMINO ACID RACEMISATION DATING (AAR)

Amino acids are the building blocks of proteins, and are present in all organisms. Amino acid dating is based on the fact amino acids start to change once an organism dies.

Amino acids can exist in two forms, which although chemically the same are mirror images of each other in terms of their molecular structure. These can be distinguished by whether they bend a beam of polarized light to the right or to the left, and are thus described as being either 'right-handed' or 'left-handed'.

In living organisms all amino acid molecules are left-handed but when the organism dies the left-handed form starts to switch to the right-handed form, in a processes called racemisation, until there are about an equal number of both types. All other things being equal, the more racemisation that has occurred, the longer the time since the organism died. If the degree of racemisation can be determined then the time that has elapsed since the organism died can be estimated. However, the age estimates provided by measuring the degree of amino acid racemisation are very general, and although we have included it as numerical dating technique here it is in many respects better thought of as a relative dating method.

In principle, amino-acid dating can be applied to a wide range of organic materials, including animal bones and teeth, plants and shells. Unfortunately there are a lot of complicating factors, which has led many to question the reliability of the technique. In recent years, however, refinements to the AAR technique, involving using protein preserved within single crystals of the calcium carbonate shells of molluscs, has led to greatly improved results.

The new AAR method limits the principal applicability of the technique to suitable mollusc shells from organic sediments. These are not found in all sand and gravel quarries but are relatively frequent and good results have been obtained from a range of Ice Age sites.

SUMMARY OF NUMERICAL DATING TECHNIQUES MOST USEFUL IN A QUARRY CONTEXT

	OSL dating	**Amino Acid dating**	**Radiocarbon dating**
Date range	Up to 400,000 BP	Up to one million years +	Up to 50,000 BP
Suitable materials	Quartz (sand) grains in section	Typically mollusc shells but also bones and teeth	Organic remains e.g. peat, plant fragments, bone and teeth
Sampling strategy	On site (directly from section)	Off site (after sieving of soil samples)	Off site (after sieving and finds identification)
What is being dated?	Time of burial, e.g. sediment deposition	Time of organism's death	Time of organism's death

CHAPTER 8: A GUIDE TO RECORDING ICE AGE SITES

INTRODUCTION: THE '4 Rs'

Important Ice Age remains can be revealed by sand and gravel quarrying as well as a range of other kinds of development that involve major ground works. This chapter contains practical advice on what should be done to ensure the recognition, recording, recovery and reporting (the '4 Rs') of such remains.

The advice is aimed at those responsible for conservation (including biodiversity and geodiversity issues) within the aggregates industry; at those responsible for advising on conservation issues within local and national government (including local government development control archaeologists); and at those engaged in fieldwork or other research relating to quarrying or other development. The latter category includes contract field archaeologists employed to undertake mitigation work.

ADVISORY AND POLICY CONTEXT

The need for the conservation of significant Pleistocene geological remains (sediments and fossils), which in most cases will only involve adequate recording and reporting rather than preservation *in situ*, and the protection of Palaeolithic archaeological remains, is covered by a number of policy and advisory documents. The two documents most relevant at the time of writing are listed below.

For the quarry industry, from the perspective of geodiversity, a key guidance booklet is *Geodiversity and the minerals industry – Conserving our geological heritage* (English Nature, Quarry Products Association and Silica and Moulding Sands Association, 2003). This booklet defines geodiversity, explains its importance, and provides practical advice on achieving conservation and good practice.

The booklet *Identifying and protecting Palaeolithic remains – archaeological guidance for planning authorities and developers* (English Heritage 1998) provides complementary guidance relating specifically to Palaeolithic archaeological remains, which are one facet of the totality of geodiversity. Both documents provide information on the legislative and policy background.

The practical advice provided below draws on these and related documents, but is informal in character and endorsement by Natural England, English Heritage, the Quarry Products Association or other relevant organisations should not be inferred. Nor should the guidance be taken to be comprehensive.

HEALTH & SAFETY AND PERMISSIONS

Working in quarries and on construction sites is potentially dangerous and Health & Safety is the top priority on these sites. All fieldwork and research must be carried out in strict accordance with Health & Safety legislation and with all rules and regulations locally in force. As a consequence, no fieldwork should be carried out without the proper and express permission of the quarry operator, contractor or other responsible organisation.

The advice given below does not cover Health & Safety, which is beyond our specific competence and needs to take account of specific circumstances. Despite this necessary omission, Health & Safety considerations should be taken as paramount in all fieldwork.

Assessment of Potential

Significant Pleistocene and Palaeolithic remains are only likely to be uncovered in a limited range of circumstances. It is always good practice to determine the potential of a proposed quarry or other development site at as early a stage as possible, ideally before extraction or other groundwork has begun. In this way, adequate mitigation measures may be put in place from the outset.

The first stage is thus a desk-based assessment of the 'Pleistocene potential', which should be carried out by suitably competent personnel. Such an assessment may be carried out as part of an environmental impact assessment, geodiversity action plan or a more wide-ranging archaeological desk-based assessment. Unfortunately, at the time of writing, not all archaeological desk-based assessments include an assessment of the Pleistocene potential of a site where this would be appropriate.

An assessment of the Pleistocene potential of a quarry or other development site will usually need to be neither elaborate nor expensive. Checking a few basic sources will usually provide most of the information needed. Some key sources are listed below.

- *Maps and memoirs of the British Geological Survey (BGS)*. These give basic data on the Quaternary history of all areas of the British Isles. Areas without Quaternary deposits on geological survey maps will be unlikely to produce Pleistocene/Palaeolithic material, although inaccuracies in published maps sometimes do allow data to be gathered from apparently unpromising locations. Significant remains are most often found in Middle or Late Pleistocene fluvial or littoral marine deposits. Aggregate deposits of glacial origin are likely to have very low potential.

- *Local government Sites and Monuments Records (SMRs) or Historic Environment Records (HERs)*. These are intended to be comprehensive computer-based records of known archaeological sites and finds, and of historic landscapes. They are the primary source of information for the provision of planning advice on archaeological issues. Unfortunately coverage of Palaeolithic period sites and finds is often patchy, and contextual information and dates may be unreliable or out of date. Furthermore, most of these records only contain information on Palaeolithic artefacts; the inclusion of information on significant fossil finds and locations has not been part of their remit. However, where significant finds have previously been recorded in the same or similar geological deposits to those under investigation, then this is of course a good indicator of the potential for future finds in these deposits.

- *Reports of the English Rivers Palaeolithic Survey*. This comprehensive survey was commissioned by English Heritage and carried out by the celebrated Palaeolithic archaeologist Dr John Wymer during the 1990s. It was initiated in specific response to "the massive increase in the quantity of gravel and sand being extracted for road building and urban development". Its aim was, amongst other things, "to identify, as accurately as possible, the find-spots of Lower and Middle Palaeolithic artefacts and the deposits containing them in order to demonstrate fully the distribution of known Palaeolithic sites in England".

    The country was divided into 12 regions for the purposes of the survey. The resulting detailed reports, each comprising a text volume and a map volume, were deposited with the appropriate SMRs/HERs, where they are available for consultation. A detailed synthesis of the results, again in the format of separate volumes for text and maps, has also been published (John

Wymer 1999 *The Lower Palaeolithic Occupation of Britain,* Wessex Archaeology and English Heritage).

- *The field guide series of the Quaternary Research Association (QRA).* These guides give up-to-date coverage of the Quaternary for many areas of Britain, including descriptions of sections and commentaries on current interpretations. Most also have reports on the Palaeolithic where finds have been made in the area covered.

These sources will often alert the reader to other potential sources for the area in question. These may include researchers and collectors who have worked in the area, together with museum and private collections.

The report of a desk-based assessment of Pleistocene potential will contain, where appropriate, an outline of the kinds of deposits and evidence which might be encountered and their potential significance (supported by maps), together with recommendations for further field-based evaluation or study.

Field-Based Evaluation

In the standard procedures for the archaeological evaluation of a potential quarrying or development site, as set out in Policy Planning Guidance Note 16 'Archaeology and Planning' (PPG 16, DoE 1990), a desk-based assessment is often followed, where justified, by a field evaluation. This may deploy a number of techniques, including systematic surface survey for artefacts contained in the ploughsoil/topsoil ('fieldwalking'), shallow geophysical survey, and the excavation of trial trenches. The purpose of this work is to attempt to determine the survival, date, character and significance of any archaeological remains that might be affected by the proposed development. This will in turn form the basis, if deemed appropriate, for a programme of archaeological mitigation work, for example formal excavation or a 'watching brief'.

Usually such field evaluations are aimed at the potential for post-Palaeolithic archaeological remains, i.e. remains from the Mesolithic through to the modern period. These remains are generally found above, on or cut into the latest deposits of sand and gravel to have been laid down during the Pleistocene epoch, or at the Pleistocene/Holocene interface. The methods employed are geared to the potential presence of such comparatively 'superficial' remains and are rarely effective for detecting the presence of Middle and Lower Palaeolithic remains, which tend to be more deeply buried within the aggregate body (although surface survey may reveal Palaeolithic artefacts).

Nevertheless, evaluation strategies should be formulated with an awareness of the potential for Pleistocene/ Palaeolithic remains where the possibility for these has been indicated by desk-based assessment. Borehole surveys, whether carried out for geotechnical or archaeological purposes, may reveal the presence of deeply buried organic-rich layers, which in the right circumstances may provide evidence for the location of deposits with good Pleistocene potential. The results of borehole surveys should therefore be exploited to help to evaluate the Pleistocene potential of a site, and in some instances it may be appropriate to carry out borehole surveys specifically for this purpose. Sufficiently detailed borehole surveys can provide valuable information on the likely depth and extent of deposits, and may be used to build up a preliminary model of deposit history.

Incorporating Pleistocene Potential into Mitigation Strategies

An assessment of the Pleistocene potential of a quarrying or development site should always form part of any archaeological or geodiversity assessment. The results of this assessment, whether or not enhanced by results from any subsequent field evaluation, need to be incorporated into the mitigation strategy that arises (if mitigation is deemed necessary). In some cases the Pleistocene potential may be the principal, or only, archaeological or geological consideration that needs to be taken account of.

In the vast majority of cases where Pleistocene potential has been identified, the most appropriate mitigation strategy will be a systematic watching brief, aimed at recognising, recording, recovering and reporting (the '4 Rs') any significant deposits and finds that are exposed during the course of aggregate extraction or other development work.

As each site is different, the specification for the watching brief (and/or other mitigation measures deemed appropriate) needs to be tailored to the specific circumstances. The specification will be based on the results of the desk-based assessment and field evaluation (if any) undertaken, the best advice available, and of course the nature of the works being undertaken (e.g. the methods and programme of aggregate extraction). Where possible, the specification for the Pleistocene watching brief should form part of the mitigation strategy included in planning permissions or site management plans. The specification should include the provision made for routine monitoring of the site (frequency and scope of routine watching brief visits) and the provision for more intensive recording and sampling in the event that particularly important deposits are uncovered.

General guidance for Pleistocene/Palaeolithic watching briefs

We noted in the previous section that watching briefs need to be tailored to the specific circumstances of each site. Nevertheless there is some general guidance that applies in most if not all circumstances, and this is provided here. The general guidance has been drawn up with the typical sand and gravel quarry in mind but parts of it can be adapted to other situations.

Personnel

Ideally, watching briefs should be carried out by an experienced Quaternary scientist or Palaeolithic archaeologist. Where this is not possible, such a person should be sought to act as the *project advisor* for the work. The project advisor will be able to call on further specialist advice and assistance when required.

Frequency of Visits

The frequency of visits will be dependent on the specific circumstances of the site, the pace of aggregate extraction and the potential of the zone(s) of the site currently under extraction. However, in general visits should be made at least once a month.

The frequency of visits can be minimised, and the effectiveness of watching briefs hugely increased, through the cooperation and involvement of quarry personnel. If quarry personnel know what to look out for (e.g. through use of this handbook) and notify researchers when anything of interest/potential interest turns up, then the benefits of quarrying to Pleistocene/Palaeolithic research can be maximised to the benefit of all.

Basic Equipment and Materials

In addition to necessary safety equipment (not covered here but including, at minimum, hard hat, high-visibility jacket and safety boots) the following items are recommended.

- Site logbook/notebook. An A4 hardback 'lab book' with alternating lined and squared pages (for notes and scale sketches respectively) is recommended.

- A ring-binder for other documentation (e.g. sediment description sheets, sample record sheets, the specification, risk assessment, etc.).

- A plan of the quarry site at a suitable scale. This should be provided by the quarry operator and will be annotated on each visit to show the progress of extraction and the approximate location of any significant deposits or finds. Multiple copies/photocopies of this may be used to prevent it from becoming too cluttered.

- A compass, hand tape and 30/50 m tape for use in on-site sketch recording. A hand-held GPS is also very useful.

- A camera, photographic scale and north point. A good digital pocket camera is adequate for most purposes.

- A range of sample bags and boxes, together with labels and markers (for finds/samples).

- A spade, trowel, spatula and brush (for cleaning of exposures and preliminary cleaning of finds).

What to Record Prior to the First Visit

Prior to the first visit some basic background information should be recorded in the notebook, such that the notebook and its purpose will make sense to a reader, otherwise uninformed, who finds it in an archive many years in the future (and will enable its return if lost). This background information should include:

- The name, address/location and OS grid reference of the quarry or site in question.

- The name, contact details and institutional affiliation (if appropriate) of the researcher.

- The name and contact details of the project advisor (if appropriate).

- The name of the quarry operating company, the quarry manager and other key quarry personnel involved, with contact details.

- Other relevant contact details (e.g. curatorial and advisory staff and organisations involved in the work; contact numbers for emergencies).

- The scope and purpose of the work. This might include reference to any relevant desk-based assessment report, evaluation report or watching brief specification, with sufficient details to allow these documents to be traced. (Copies of these documents should be included in the ring binder.)

- Details of the site code to be used to identify any finds collected, and to identify drawings, photographs and other supplementary documentation.

- Health & Safety information, such as risk assessments and the rules and regulations applying on the site.

What to Record on Each Visit

In quarry sections (working faces) or other deep excavations the visibility of Palaeolithic and Quaternary material may not be high, although large mammalian bones are unmistakeable and are usually found with other, less visible fossil and archaeological finds. In most fluvial contexts, river terraces are built on cold stage gravel, which usually contains only re-worked bones or archaeology. Most *in situ* remains occur at the base of gravel faces or interbedded between gravel units. Any watching brief therefore requires careful monitoring of the basal layers of sand and gravel sections. Significant remains may also be found in other sediment traps, such as lake sediments or raised beaches.

Organic remains capable of providing environmental information are in most cases found as the basal layers of terrace gravels. In general they are recognised by their fine grain size (muds, sands) compared to the gravels of the terrace. Organic sediments often contain wood, leaves or seeds, and mollusca, micro-vertebrates and insects. These may be easily visible in sections if present in large quantities. Any fine-grained sediments warrant closer attention if they contain visible remains, and when sieved in the laboratory may even yield significant material that is not visible in field sections.

At least three tasks should be undertaken on each visit:

- The quarry manager and other appropriate quarry personnel should be interviewed. Any information on finds and potentially significant deposits uncovered should be noted. Any finds recovered by quarry staff should be listed, described and located to their place and deposit of origin as best as is possible (using sketch maps as appropriate), but making clear the level of confidence involved. They should be bagged and labelled adequately, so that cross-referencing between the finds themselves and the list recording them is always unambiguous. Information should be obtained on the intentions for extraction in the immediate future (such as may guide the appropriate timing of future visits).

- Exposed sections (working faces) and the quarry floor should be carefully examined for finds and deposits of interest. As accurately as is practical, the position of working faces at the time of the visit should be marked on the site plan (the general idea is that it should be possible to reconstruct, from the records of each visit, the progress of ex-

Fig. 147: Searching waste heaps for Palaeolithic handaxes in gravel pits at Dunbridge, Hampshire. Over 1,000 handaxes have been found at this quarry.

traction during the course of the watching brief). A note should be made of areas of floors and faces that are obscured and of areas where access was not possible for Health & Safety or other reasons. The weather conditions at the time of inspection should be noted. An attempt should be made to interpret exposed sections in terms of the geological succession understood for the site and representative photographs (with scale) taken. Finds and deposits of particular significance (e.g. organic-rich horizons) should be recorded in more detail (as described in the next section).

- The waste heaps beside the quarry graders at the gravel screening plant should be inspected and any fossil material or artefacts recovered and recorded. Much large mammal material and larger artefacts such as handaxes are rejected by the gravel screening plant and end up on the waste heaps beside the quarry graders (Fig. 147). As these waste piles are removed systematically every few weeks by most operators it is often possible to determine from which general area of the quarry they derive and their probable geological context (material adhering to finds may help here). This alone may be a significant prospecting tool if bones or artefacts are in low concentration in sands and gravel. A note should be made of the state of the heaps (such as may affect the possibility of recognising material) and the intensity of the inspection which was achieved.

Each quarry visit should be properly logged, even if the outcome is entirely negative.

What To Do if Significant Finds or Deposits are Encountered

Judgment must be exercised as to what represents a 'significant find or deposit' and what level of response is appropriate. The following is intended only as general guidance. It should be read in conjunction with chapters 4–7, which describe in more detail the procedures to be followed with respect to sediments, fossils, artefacts and dating.

Isolated finds of larger artefacts (e.g. handaxes) and mammal bones and teeth may be found *in situ* in quarry faces or apparently dislodged from their context. In the former case it is usually sufficient to record the location and stratigraphic context of the find, supported by measured sketches and photographs. In the latter case an attempt should be made to identify the deposit from which the find is likely to have been dislodged, with an indication of the confidence of this identification.

The recording of isolated finds in this way, when repeated over numerous visits as extraction proceeds, allows an archive of find spots in three dimensions to be built up, which can be used to reconstruct aspects of the depositional environment.

If extensive fossil-bearing sediments are encountered, these will require systematic sampling for specialist analysis. At this point a decision needs to be made based on whether the exposure can be protected from further disturbance until sampling by specialists can be organised. This will require negotiation with the quarry manager, and amongst other things it will be dependent upon whether aggregate extraction can proceed satisfactorily leaving the fossil bearing sediments, or a representative portion of them, intact until specialist investigation and sampling can take place. Watching brief specifications should include provision for this occurrence.

If it *is not possible* for the sediments to be protected for specialist investigation and sampling, then the researcher should undertake what recording and sampling are practicable within the time and resources available. This may include the following:

- Making measured sketches and taking photographs of the sediments and stratigraphic sequence. The location should be fixed as accurately as possible.

- Formal description of the sediment sequence. Guidance on this is provided in chapter 4 and an example of a *sediment description sheet* is provided as Appendix 1.

- Remove (and bag and label) larger visible fossils and artefacts. The extent of excavation will be limited by time and safety considerations.

- Take bulk samples (ideally 20 litres) for the potential recovery of vertebrate microfossils, molluscs, insects, plant macrofossils etc.. Although the samples will contain a mixture of material, at least one sample should be taken for each fossil category. If in doubt take more samples than might ultimately be needed – unwanted samples can always be thrown away. Samples should be recorded using *sample record sheets* (an example is provided as Appendix 2) and the precise position from which each sample was taken should be marked on the measured sketch of the stratigraphic sequence. If the exposed section includes a stratigraphic sequence of fossil bearing deposits, samples should be taken from each stratum, with the sequence of samples clearly recorded.

If it *is possible* for the sediments to be protected for specialist investigation and sampling, then the first two stages in the list above – illustration and sediment description – should be carried out immediately, and any loose bones or artefacts should be removed, bagged and labelled, with positions shown on the sketch section. Where the deposits are extensive enough, bulk samples should be taken as described above as an 'insurance policy'. Misunderstandings and accidents do occur and the researcher should use their judgement as to what to finds and material should be removed immediately to 'be on the safe side' and what should be left for specialist investigation at a later date.

PRIMARY CONTEXT DEPOSITS, HUMAN REMAINS AND CUT-MARKS

In all cases the researcher should be alert to the possibility of primary context material – that is, material that has been undisturbed or only little disturbed since the time of its deposition. Such contexts are rare and of exceptional importance. They will almost invariably be associated with relatively fine sediments, indicating deposition in a low energy environment. The following are indicators of primary context deposition:

- Well-preserved mammal bones in articulation or partial articulation, or bones apparently representing several elements of a single skeleton.

- Lithic artefacts in 'fresh' condition, with little or no evidence of damage by transport ('rolling').

- Lithic assemblages including fine debris (under c.3 mm) from knapping.

- Lithic assemblages with a high flake/blade to core/core tool ratio (commonly greater than 100:1). Natural disturbance tends to lower this ratio very substantially.

- Lithic assemblages with conjoinable pieces. These are very difficult to recognise but may include breaks or 'technological conjoins' (where two or more pieces from a knapping sequence can be fitted together).

As was noted in chapters 3 and 5, human remains of Pleistocene date are extremely rare finds and any such discovery is likely to be of international significance. Therefore, one must always be especially alert to the possibility. When unarticulated, to the untrained eye much of the post-cranial (below the neck) human skeleton is not particularly recognisable. However, the human skull is distinctive, even when fragmentary, as are human teeth. Teeth are also the most durable part of the skeleton and therefore the part that survives best.

The context in which human remains or suspected human remains are found is very important for their correct interpretation. Later prehistoric and medieval human burials are a relatively common feature on gravel sites. They will be found within the body of the sands and gravels because the grave was dug down into these; the grave cut should be recognisable. (Animal burials, for example those made during foot-and-mouth outbreaks, are similarly found in graves dug into the aggregate body.) Not until towards the end of the Pleistocene, however, were human remains sometimes given formal burial. Therefore, in terms of what happened to the body after death (scavenging by animals etc.) and how the bones of the skeleton came to be scattered and perhaps incorporated into geological deposits as fossils, the processes involved with most human remains from the Pleistocene are the same as those for other mammals of similar size.

By far the most common evidence of human activity is provided by stone tools. However, another important type of evidence of human activity is provided by the cut marks made by stone tools which are sometimes found on animal bones (see chapter 5). Researchers should be alert to the possibility of finding this evidence of butchery.

Reporting

On the completion of a watching brief or other fieldwork a report or reports must always be prepared. The minimum requirement is for a report which will be lodged in the appropriate local authority Sites and Monuments Record or Historic Environment Record. This report will include the following:

- Site name, location and extent, including the Ordnance Survey grid reference (supported by appropriate maps and plans).

- The background to the work – why and when it was carried out, by whom, etc..

- The methodology – details of any method statement or specification for the work should be included.

- An outline of results (supported by appropriate drawings and photographs).

- An interpretation of the results and assessment of their significance.

- Recommendations for any further fieldwork or analysis deemed appropriate.

A short report should be prepared even if the results of the watching brief were entirely negative. This is important because the report will provide the only formal documentation that the work was actually carried out. Furthermore, the report will provide an assessment of *why* the results were negative (e.g. because significant Pleistocene material was indeed absent or because of limitations, practical or otherwise, to the conduct of the work). This will be important in guiding decisions concerning requirements for future watching briefs in related situations.

If the work has produced results of sufficient scientific importance, then a report or reports should also be prepared for publication. This is likely to be the case when the watching brief led to detailed specialist investigation and analysis. The nature of such a report or reports produced will depend on the importance of the results and the discipline(s) to which they are most relevant. For example, results of mainly archaeological significance but of primarily of local or regional interest may be published in regional archaeological journals, e.g. the journals of county archaeological societies. Where the work is of national significance publication in a national archaeological journal, such as the *Proceedings of the Prehistoric Society* should be considered. *Lithics*, the journal of the Lithics Study Society, takes papers primarily focussing on stone tools and their interpretation.

Where the results are of mainly geological significance, then publication in regional or national geological journals should be considered, e.g. the *Proceedings of the Geologists' Association*. Work of major importance and interdisciplinary in scope may be published in the publications of the Quaternary Research Association, the *Quaternary Newsletter* (for shorter pieces) and the *Journal of Quaternary Science* (for major papers). These examples provide only a sample of the range of publication vehicles available, the choice of which will depend on nature and overall significance of the results.

Deposition of Finds and Records

Following the completion of fieldwork and analysis, arrangements should be made for the deposition and long-term conservation of finds (both artefacts and fossils) and records in a suitable repository, usually a regional or national museum. Advice on this should be sought at the earliest possible stage and ideally arrangements for the ultimate deposition of finds should be included in the specification or method statement prepared in advance of the work (but only following consultation with the proposed receiving body). Conservation of delicate fossil remains can be expensive but unfortunately it is impossible to determine what the requirements for such conservation might be prior to the actual discovery of the material.

Glossary: An Ice Age Dictionary

Introduction

The aim of this glossary is to provide definitions and explanations of technical terms and techniques relating to the Ice Age that the reader is likely to encounter in wider reading, and not just those terms used in this book. Quaternary science covers many disciplines, and the entries in the glossary are drawn mainly from the fields of Palaeolithic archaeology (especially terms relating to stone tools), palaeoanthropology, geology, geomorphology, palaeoclimatology and the biological sciences (especially palaeontology). We have been selective, and while, for example, there are quite a number of terms that relate to glacial landforms, which are central to the theme of the book, there are much fewer that relate to the processes and forms of glaciers themselves. We have included a few terms, mostly the names of rocks and periods, relating to geological periods earlier than the Quaternary, because these earlier rocks, transported by glaciers and rivers, form much of the content of Ice Age deposits, or were a resource used for tool making by prehistoric peoples.

We have not provided any entries for particular archaeological or geological sites, although these are mentioned when they form the type site for archaeological industries or geological units and events. We have, however, included entries on some important geographical features, such as the Bytham River, or geological formations, such as the Cromer Forest-bed Formation. In both archaeology and geology there is a proliferation of regional terms: emphasis in this glossary is on Britain first and continental Europe second, with regional terms for other parts of the world almost entirely omitted.

Some very basic terms are included, because these terms may not be understood by somebody unfamiliar with the particular discipline in question. Furthermore there are many words that have one meaning in everyday usage but have a very specific and sometimes quite different meaning in a specialist discipline; the everyday meaning can be misleading. For example 'blade', when used of stone tools, 'raft' when used in geology, and 'family' when used in biological taxonomy, all have very specific, non-obvious meanings. We have included a few entries on Ice Age animals that may be less familiar, such as the musk ox, but have not included well-known ones, such as the mammoth.

Conventions Used in Glossary

There is extensive cross-referencing between the entries, and when terms found within an entry are written in italics, this means that there is a separate entry for that term elsewhere in the glossary. The normal convention is that the scientific names of biological species are written in italics, but because italics is used here to denote cross-referencing these names are underlined instead. Where dates are given these are indicative only, but they relate to a time-scale measured in 'real' calendar years. Most archaeological dates within the Upper Palaeolithic period are based on radiocarbon dating, and many archaeologists prefer to use a time-scale for this period based on uncalibrated radiocarbon dates because there is at present no agreed calibration. Largely for the purpose of consistency we have chosen to calibrate all dates, despite the acknowledged difficulties and dangers. This means that the dates that are found in much of the literature for a particular finds, find types or events will be as much as 3,000 years younger than the dates tentatively given in this glossary (and throughout the book) for the same finds and events.

AAR: see *amino acid racemisation.*

Abbevillian: a name given to the supposed earliest (i.e. pre-*Acheulean*) *handaxe industries* of Europe, characterised by crudely chipped handaxes. As it is now known that finely-made handaxes may also occur at a very early date in Europe (e.g. at Boxgrove about 500,000 years ago), the reality of the Abbevillian as an industry distinct from the Achuelean is now questioned, and the term has largely fallen out of use. Named after Abbeville on the Somme in France, where handaxes were found in gravel pits in the 19th century.

Ablation: the loss of mass from a *glacier*, largely through melting.

Ablation till: debris (*till*) deposited by a *glacier* as a result of the melting away of the ice around it. The *clasts* in ablation till tend to more angular than those found in *lodgement till*.

Absolute dating: techniques of dating which provide an estimate of age in calendar years, i.e. in years before present or *BP*. Alternative terms are *chronometric dating, numerical dating* and *sidereal dating*. See also *radiometric dating*. Compare *relative dating*.

Accelerator Mass Spectrometry (AMS): in *radiocarbon dating,* a technique that allows very small pieces of bone and other organic materials to be dated. In contrast to the conventional radiocarbon method, in AMS the proportion of the different carbon *isotopes* in the sample is determined by direct measurement of the weight of the atoms, disregarding their radioactivity.

Acheulean: a *Lower Palaeolithic* stone tool *industry,* characterised by pebble tools, *bifacial* tools (including *handaxes*) and *flake* tools (*Mode 2*). Named after the *type site* of Saint Acheul in France.

Active layer: the upper layer of the ground in *permafrost* regions, which freezes in the winter and thaws in the summer.

Adaptation: any aspect of an organism's anatomy, physiology, or behaviour that evolved to perform a particular function.

Aeolian deposit: a deposit which was transported and deposited by the wind. See also *coversand* and *loess.*

Aerobic: a term that describes conditions in which oxygen is present, causing disintegration of organic remains such as plants and the soft parts of animals. Opposite of *anaerobic*.

Aerosol: a suspension of fine solid particles or liquid droplets, other than ice or water, in the atmosphere. Aerosols resulting from volcanic eruptions can cause cooling at the Earth's surface.

Aggradation: in *river terrace* formation, aggradation is the deposition of material by the river, causing the build up of deposits in the river valley. Compare *downcutting*.

Aggregate: literally a collection of different things; in the context of this book aggregate refers to sand and gravel deposits laid down by rivers and glaciers.

Ahrensburgian: the name given to a *Terminal Upper Palaeolithic* industry, of about the time of the *Younger Dryas Stadial* (c.12,900–11,400 BP), found principally in northern Germany and the Low Countires. Ahrensburgian points (characteristic small *tanged* points) have also been found in southeast England, suggesting that the Ahrensburgian industry spread as far as Britain (i.e. western *Doggerland*). Named after the *type site* of Ahrensburg, near Hamburg.

Albedo: the albedo of a surface is a percentage measure of the amount of incoming solar radiation that it reflects. Areas covered by snow and ice are highly reflective and thus have a very high albedo, up to 90%. Areas covered by forest, on the other hand, have low albedo.

Allerød Interstadial: the name given in continental northwest Europe to an *interstadial* (c.13,800- 12,900 BP) during the *Lateglacial*, which also represents the last major warm period before the present *Holocene interglacial*.

Preceded by the *Older Dryas Stadial* and followed by the *Younger Dryas Stadial*. Equivalent to the latter part of the *Windermere Interstadial* in Britain. See also *Bølling-Allerød Interstadial*. Named after the *type locality* in Denmark.

Alluvium: material deposited by rivers. Although *Pleistocene* sands and gravels are by this definition alluvium, the term is sometimes restricted young deposits of *Holocene* age (e.g. fine sands and silts).

Alternate flaking: in stone-tool manufacture, the working of a *core* or blank by removing *flakes* alternately from either side of the piece.

Amino acid dating: a means of *absolute dating* based on systematic changes in amino acids (the building blocks of protein) after the death of an organism, a process called *amino acid racemisation*. As this happens at a known rate (other things being equal) the amount of change can be used to determine how much time has passed since death. The technique can provide dates of up to a million years old, and is typically undertaken on *mollusc* shells.

Amino acid racemisation (AAR) dating: see *amino acid dating*.

AMS: see *Accelerator Mass Spectrometry*.

Anaerobic: a term which describes conditions in which oxygen is absent, slowing or halting the disintegration of plant and animal material. Opposite of *aerobic*.

Ancaster River: the name given to an ancient river system believed to have drained the southern Pennines before the *Anglian Glaciation*, flowing through the Ancaster Gap into northern East Anglia. Compare *Bytham River*.

Andesite: a type of rock of volcanic origin, named after the volcanoes of the Andes, occasionally used for making stone tools (including *handaxes*) during the *Lower Palaeolithic*. The fine-grained quality of the rock made it a suitable substitute for *flint* where the latter was absent. Outcrops in Britain include the Lake District and north Wales.

Anglian Glaciation: a major *glaciation* that occurred between about 474,000 to 427,000 years ago (*MIS* 12), characterised by an ice sheet that covered much of Britain and reached as far south as London. Named after East Anglia, where physical traces of this glaciation are particularly apparent. Equivalent to the *Elsterian* glaciation of continental northwest Europe and the *Mindel* glaciation of the Alps.

Anthropogenic: caused by human action (e.g. anthropogenic climate change).

Anthropology: literally 'the study of man'. Social or cultural anthropology is mainly the study of present-day non-western or 'traditional' peoples and societies. See also *physical anthropology* and *palaeolanthropology*. Compare *archaeology*.

AP: in *pollen diagrams*, an abbreviation for 'arboreal pollen', i.e. all tree pollen taken together as a group. Compare *NAP*.

Archaeology: the study of past human behaviour and societies by analysing the material remains (sites and artefacts) that survive today. Compare *anthropology, palaeoanthropology, palaeontology* and *geology*.

Artefact: any product of human workmanship.

Articulated: with reference to bones, describes situations in which *fossil* bones are found lying in the same relationship to one another as they had in the skeleton during life. When articulated bones are found in Ice Age deposits (which is rare) it is assumed that they reached their final resting place while flesh and ligaments still held them together. Opposite of *disarticulated*.

Artiodactyla: the *order* of *mammals* comprising the even-toed *ungulates* (i.e. cloven-hoofed animals). Examples include pigs, deer and cattle (compare *Perissodactyla*).

Assemblage: a group of different items (*artefacts* or *fossils*) found in association with one another at the same site.

Astronomical theory: in *Quaternary science*, the theory that *glacial–interglacial* cycles are driven by cyclical changes in the Earth's orbit, tilt and spin with respect to the Sun.

Atlatl: spear-thrower.

Auger: a device used to bore through sediments to test their nature or extract sediment cores.

Aurignacian: one of the earliest *Upper Palaeolithic* (*Mode* 4) *industries* found in Europe, named after the French *type site* of Aurignac. Dated from before 40,000 *BP* to about 32,000 BP. Characterised by a diverse range of tools made of stone, bone and antler, and associated with the earliest unequivocal art objects (e.g. carved and engraved animal figurines made of ivory and bone) found in Europe.

Aurochs (plural **Aurochsen**): extinct wild cattle, Bos primigenius (*Order Artiodactyla; Family* Bovidae, *bovids*). Common during the Ice Age, the last aurochs died in 1627 in the Jaktorow Forest near Warsaw.

Autocthonous: another word for indigenous; used in *geology* and related sciences to mean 'formed in the place where it is found'.

Aveley Interglacial: the name sometimes given in Britain to the *interglacial* corresponding to *MIS 7* (about 242,000–186,000 *BP*). Also called the Aveley/West Thurrock Interglacial. Named after the *type site(s)* on the Lower Thames where deposits believed to correlate with this interglacial have been studied.

Azilian: a *Final Upper Palaeolithic* or *Epipalaeolithic* stone-tool industry found in southern France and Iberia. Named after the *type site* of Le Mas d'Azil in the French Pyrenees. See also *Federmesser*.

Backed: in stone-tool making, a backed tool (typically a *blade*) is one that has been deliberately blunted along the side opposite the cutting edge by means of abrupt vertical *retouch*. In some cases the purpose of this may be to protect the hand while using the tool, while in others it may be to facilitate *hafting*.

Bar: in *sedimentology*, an elongated deposit of sand, shingle or mud in a river channel or in the near-shore part of the sea or a lake.

Barbed point: a harpoon-like projectile point, made of bone or antler, with a row of barbs running down one side (a uniserial barbed point) or both (biserial). A harpoon point proper will have a hole or *tang* for the attachment of a line. Found in the *Late Upper Palaeolithic* as well as the *Mesolithic*.

Basalt: a fine-grained *igneous* rock. Extensively used for stone-tool making in Africa.

Bas-relief: a sculpted figure on a surface (e.g. a cave wall) produced by removing the material around it so that the figure stands out.

Baton: the name giving to an object of unknown use found on *Upper Palaeolithic* sites. It consists of an antler rod with a hole through the thickest part of the head. From the French, bâton de commandement, which suggests a ceremonial use. A more prosaic, functional interpretation is that these artefacts were used as *shaft-wrenches*, a device for straightening the shafts of spears (however, some examples from France are highly decorated).

Bec: in stone-tool typology, a type of borer.

Bed: in *lithostratigraphy*, a well-defined layer (stratum) in a stratigraphic section believed to have originated in a single depositional event (albeit perhaps of long duration). In geological nomenclature, a bed is the smallest stratigraphic unit that can be given a formal name, e.g. the West Runton Freshwater Bed. Beds form part of *Members* or *Formations* (the main stratigraphic unit), e.g. the West Runton Freshwater Bed forms part of the *Cromer Forest-bed Formation*.

Bedded deposit: in *lithostratigraphy*, a stratified deposit with well-defined layers (strata).

Beestonian: the name given in Britain to the *Early Pleistocene* stage, approaching a million years in duration, preceding the *Cromerian* (sensu lato or *Cromer Complex*). Like the Cromer Complex it consists of numerous alternating *glacial* and *interglacial* phases, from *MIS* 60(?) to MIS 22, about 1.8 *mya* to 865 *kya*. Named after Beeston Cliffs near West Runton in Norfolk, where deposits of this stage are preserved.

Benthic organisms: organisms that live on the sea floor.

Biface: a sharp-edged implement, with both faces flaked. An alternative name for *handaxes* and related stone tools. Bifaces are the defining element of *Mode 2 industries.*

Bifacial: with reference to stone tools, bifacial means worked on both faces (for example a *handaxe*).

Biome: a term used to describe an ecological community of plants and animals established over a wide area.

Biostratigraphic dating: a technique of dating deposits by means of the *assemblages* of plant and animal *fossils* found within them, which differ from period to period. Furthermore, many *mammalian* species underwent evolutionary change or extinction during the course of the *Pleistocene*, and the presence/absence of particular species also allows deposits to be dated. See also *Mammal Assemblage Zone (MAZ), Vole Clock* and *Pollen Zone*.

Biostratigraphy: a stratigraphy based on the nature of the *fossils* found in deposits and the changes in animal and plant life through time. See also *Mammal Assemblage Zone* and *Pollen Zone*.

Bioturbation: disturbance of sediments, including burrowing, caused by the activity of animals living in the sediment.

Bipedal: walking on two legs. Compare *quadrupedal*.

Bipolar core: with reference to stone tools, a term usually applied to *blade cores* that have a *striking platform* at both opposing ends.

Bipolar technique: in stone-tool making, a method of flaking in which one end of the *core* is placed on an anvil while the other end is struck with a *hammer*, such that *flakes* are detached simultaneously from both ends when the core is struck.

Bison: a humpbacked, shaggy-haired wild ox (*Order Artiodactyla; Family* Bovidae, *bovids*). Only two species survive today, the American bison, Bison bison, (commonly called the 'buffalo'), and the European bison, Bison bonasus (also called the *wisent*). Several Ice Age species are recognised, of which the most common, the steppe bison (Bison priscus), which was frequently depicted in Ice Age art, became extinct at the end of the *Pleistocene*.

Bivalve: members of the Bivalvia *class* of the Mollusca (*mollusc*) *phylum*, having a two-part shell. Exclusively aquatic, over 30,000 species of bivalve have been identified, including scallops, clams and oysters.

Blade: in stone-tool typology, technically any *flake* more than twice as long as it is wide can be classified as a blade. However the term is more usually applied to systematic production of parallel-sided flakes, with one or more *dorsal* ridges and more than 2.5 cm long, from specially prepared *cores*, seen in the *Upper Palaeolithic*.

Bladelet: in stone-tool typology, the same as a *blade* but less than 2.5 cm long.

Bølling-Allerød Interstadial: in continental northwest Europe, the name given to the *Bølling* and *Allerød Interstadials* combined, interrupted by the *Older Dryas Stadial*. Equivalent to the *Windermere Interstadial* in Britain and sometimes called by the general name of the *Lateglacial Interstadial*. From about 15,400 *BP* to about 12,900 BP.

Bølling Interstadial: in continental northwest Europe, the name given to the *interstadial* following the *Last Glacial Maximum* and preceding the *Older Dryas Stadial*. Dated from approximately 15,400–14,300 *BP* and corresponding to the first part of the *Lateglacial Interglacial* (*Windermere Interstadial* in Britain). See also *Bølling-Allerød Interstadial*.

Boreal: northern.

Boulder clay: an alternative, British and somewhat old-fashioned term for glacial *till*.

Bout coupé: a type of small, flat-based, cordiform (heart-shaped) *Mousterian handaxe*. In Britain they are associated with the *Neanderthal* reoccupation of Britain around 60,000 *BP,* following a long period of apparent human absence. See also *Mousterian of the Achulean Tradition* (*MAT*).

Bovid: a bovid is any one of numerous cloven-hoofed species within the *Family* Bovidae (*Order Artiodactyla*). Examples include cattle, *bison,* antelopes, sheep and the *musk ox*.

BP: abbreviation for 'Before Present'.

Braided river: a river which flows in a net of shifting channels, due to more material being brought into the river than it is able to transport. This type of *fluvial* system is characteristic of cold climates, was common during the Ice Age, and can be seen in Arctic environments today.

Breccia: a deposit, typically found in limestone caves, consisting of angular rock fragments consolidated in a *calcitic* maxtrix. It is formed when the rock is detached from the cave wall through frost action or alternating heat and cold.

Brickearth: in southeast England, the name given to silty clay deposits that are suitable for making house bricks, and have been extensively quarried for this purpose (e.g. in the Thames Valley). They may have a *loess* component but are water lain.

Browser: a *herbivore* that eats predominantly the leaves of trees and shrubs.

Brunhes Chron: the present period of normal geomagnetic polarity (i.e. with magnetic north situated near the North Pole), which began about 780,000 years ago. The beginning of the Brunhes Chron is taken to mark the transition from the *Early Pleistocene* to the *Middle Pleistocene*. Named after Bernard Brunhes (1867-1910), one of the scientists who discovered the phenomenon of *geomagnetic reversals*. Compare *Matuyama Chron*.

Brunhes-Matuyama Reversal: the name given to the *geomagnetic reversal* event that took place about 780,000 *BP.* It marks the transition between the *Matuyama Chron* and the *Brunhes Chron,* and also between the *Lower Pleistocene* and the *Middle Pleistocene*.

Bulb of percussion: in stone-tool manufacture, a pronounced swelling on a detached *flake*, just below the impact location of the *hammer* strike. A bulb of percussion is rarely present where flakes have been produced naturally, for example by impact between pebbles in a river or by frost fracture. The presence of a bulb of percussion is therefore a useful indicator of deliberately struck stone. Hollows, called *negative bulbs of percussion*, may be present on the *core* from which flakes have been struck.

Bulbar scar: an irregularly shaped scar found on a *bulb of percussion* marking the place where a small piece of stone has been detached during fracture. Common where impact has been maximised (e.g. a *hard hammer* has been used).

Bulk sediment sampling: the removal of quantities of sediment large enough to contain representative samples of evidence such as plant and animal *microfossils*. In *Ice Age* contexts these samples tend to be approximately 20 litres.

Burin: a stone tool with a chisel-like edge, made by striking off a thin spall down the side of a *flake* or *blade*. Typically associated with the carving of bone, antler or ivory.

Butt: in stone-tool typology, a butt on a *flake* is the remnant of the flat surface (*striking platform*) from the *core* from which the flake was struck.

Bytham River: a major ancient river that used to flow through the Midlands into East Anglia, and appears to have been a focus for *hominin* colonisation and occupation. It was obliterated by the *Anglian* ice sheet around 450,000 *BP* although alluvial deposits, containing *fossils* and *artefacts*, survive to chart its course and are commercially quarried.

Calcareous: of rocks or sediments, containing calcium carbonate (mainly from shells); chalky.

Calcite: a mineral of calcium carbonate. Often the primary constituent of the shells of marine organisms such as *coccoliths* and *foraminifera*, and consequently a common constituent of *sedimentary rocks*, such as *limestone* and *chalk*, largely formed from the shells of dead marine organisms.

Calibration: principally in *radiocarbon dating*, the conversion of 'raw' dates into 'real' calendar years. Due to changes in the concentration of radiocarbon (^{14}C) in the atmosphere over time, unless they are calibrated the dates provided by the radiocarbon method do not correspond to calendar dates but deviate from these, usually being too young. The main method of calibrating radiocarbon dates is by obtaining dates from tree rings, the exact calendar date of which is known, and building up a calibration curve by means of which all dates falling within the range of *dendrochronology* can be corrected. Unfortunately tree-ring dates can only provide reliable calibration back to around 11,000 BP. The period from then up to the effective limit of the radiocarbon method (for reason of the relatively short half-life of the ^{14}C *isotope*), around 50,000 BP (which corresponds to the *Upper Palaeolithic* period in Europe), cannot be calibrated using tree rings. Other, less reliable methods have to be used, including lake and marine *varves*, and *Uranium-series* dates on corals. Although there is currently no agreed calibration of radiocarbon dates in this range, calibration is important if archaeological dating is to be brought into line with palaeoclimatic dating derived from *ice cores*, etc., which provide real calendar dates. In this book and glossary, all dates are presented in terms of calendar years, even if this can only be indicative, and all dates derived ultimately from radiocarbon determinations have been calibrated using a computer program called *CalPal*. As much of the literature on the Upper Palaeolithic uses uncalibrated dates, there will often be a major discrepancy between the indicative dates given in this book and those to be found in that literature. For example, between 18,000 BP and 40,000 BP radiocarbon dates are about 3,000 years too young.

CalPal: abbreviation for the 'Cologne Radiocarbon CALibration & PALaeoclimate Research Package'. A computer program package that provides calibrated *radiocarbon dates* for the full range of the radiocarbon dating technique back to 50,000 *BP*. The package uses a range of methods for *calibration* of dates beyond the limits of *dendrochronology* and is somewhat controversial.

Calving: where glaciers border the sea, the process whereby fragments of the glacier break off to form icebergs is known as calving.

Cambrian: the geological *period* from about 542–488 million years ago.

Canine tooth: a pointed tooth between the *incisors* and *premolars*, often greatly enlarged in *carnivores*.

Cannon bone: the third *metacarpal* (or shin bone) of a horse, which provides the major support for the body weight.

Caprids: mammals belonging to the sub-family Caprinae (*Family* Bovidae; *Order* Artiodatctyla). Mostly used of sheep and goats (including ibex) but, surprisingly, the *musk ox* is also a caprid and more closely related to sheep than oxen.

Carboniferous: the geological *period* lasting between about 359 and 299 million years ago. The name means 'coal-bearing' and *glacial till* incorporating Carboniferous rocks is often black in colour.

Carnassial teeth: the first lower *molar* and fourth upper *premolar* in *carnivores*, in which these teeth have evolved blade-like structures for shearing meat.

Carnivore: an animal adapted to a meat-based diet, readily identifiable by specialised *canine* and *carnassial* teeth.

Carpal bones: the wrist bones in a human, or the equivalent bones in the forelimb of other *mammals*.

Caudal vertebrae: the *vertebrae* in the tails of *mammals*.

Cenozoic: the present geological *era*, which began with the extinction of the dinosaurs about 65 milllion years ago. The evolution of the *primate order* of *mammals* took place during this era. Cenozoic means 'recent life'.

Cervical vertebrae: the *vertebrae* of the neck. Almost all *mammals*, including giraffes, have seven cervical vertebrae.

Cervid: animals of the deer family (*Family* Cervidae; *Order* Artiodatctyla).

Chaîne opératoire: in stone-tool making, this term describes the whole sequence of actions from the initial selection of raw material, through the *reduction* of a *core*, to the discard of exhausted tools.

Chalk: a soft white *sedimentary limestone*, created as seafloor deposits during the *Cretaceous period* and containing nodules of *flint*.

Channel River: the name given to a massive ancient river which, prior to the flooding of the English Channel, flowed along the course of the channel out into the Atlantic. Its tributaries included the Thames, Rhine, Somme and the ancient Solent River.

Chatter-marks: characteristic pock-marks on pebbles caused by agitation of the stones against one another in water, e.g. through tidal action.

Cheddar point: a type of trapezoidal *backed blade* of the *Late Upper Palaeolithic Creswellian industry* (*Mode 4*).

Chert: fine-grained rock suitable for stone-tool making but not as good quality as *flint* (which is a very fine-grained form of chert). Occurs as siliceous nodules or sheets in *sedimentary rocks* such as *limestone*. See also *Greensand chert*.

Chironomidae: non-biting midges. Fossil remains of chironomids may be used in environmental and climate reconstruction, although the technique is not as developed as that using *Coleoptera* (beetles).

Chopper: in stone-tool technology, a simple tool with a sharp cutting edge produced by the removal of *flakes* from one side of a *core*.

Chronometric dating: see *absolute dating*.

Clactonian: a British *Lower Palaeolithic* stone-tool *industry* characterised by pebble *cores*, *flake* tools and an absence of *handaxes* (*Mode 1*). The name derives from the *type site* of Clacton-on-Sea, Essex.

Clade: in evolutionary biology, a group sharing a single common ancestor – i.e. a single evolutionary branch.

Class: in the *Linnaean system* of classifying organisms, class is the rank in the hierarchy below *phylum* and above *order*. For example, <u>Homo</u> <u>sapiens</u> belongs to the chordate phylum, the *mammalian* class and the *primate* order.

Clast: a rock fragment.

Clastic: consisting of or containing rock fragments.

Cleaver: in stone-tool technology, a *biface* or *handaxe* with a wide axe-like cutting blade at the tip (instead of a point), generally at approximately right-angles to the long axis of the tool.

Climate: the long-term statistical average of weather conditions. The parameters of climate include not just temperature but also air-pressure, precipitation, soil moisture, runoff, cloudiness, storm activity, winds and ocean currents.

Climatic optimum: a phase during an *interglacial* when temperatures are significantly higher than the long-term average for the period as a whole.

Climax community: in ecology, the term used to describe a community of plants and animals that has achieved a 'steady state' balance with its environment and will not change further without change – usually climatic change – to that environment. The idea is that when the climate of a region changes, plants and animals will respond to this change at different speeds and in different ways, giving rise to what is known as an *ecological succession*. This process will end with a climax community, representing the community of plants and animals best adapted to the average conditions in that region. This is a useful concept, even if much criticised as an oversimplification of the complexities of ecological and evolutionary reality.

Clinometer: in *geology* and *sedimentology*, an instrument used to measure the inclination of *beds*, the axes of *clasts*, etc..

Coccoliths: very small *calcite* plates that are formed on the surface of flagellate organisms. The most common calcareous *microfossils*.

Coleoptera: the *order* within the *class* Insecta commonly called beetles (*Order* Coleoptera; *Class* Insecta). There are about 350,000 described species of beetles; many more are yet to be described. Often very specific in their habitats and climate tolerance, beetles provide excellent *proxies* for environmental and climate reconstruction. Coleoptera means 'sheathed wing'.

Colluvium: weathered material transported downslope by gravitational forces. Another term for *slopewash*. Compare *alluvium*.

Column sampling: the removal of a single large sediment sample or series of smaller samples through a vertical sequence of sediments, allowing any changes in *microfossil* content through time to be identified. See also *monolith tin*.

Conchoidal fracture: the way that *flint* (and glass) fracture when struck, with a smooth, curved (literally 'shell-like') surface.

Cone of percussion: in stone-tool manufacture, a cone-shaped feature that forms under the *point of percussion* on the *butt* of a *flake*.

Conformable: in geology, a term used when a *bed*, or larger stratigraphic unit, lies parallel with the bed or surface on which it rests. Compare *unconformity*.

Continental drift: the lateral movement of continents as a result of sea-floor spreading. See also *plate tectonics*.

Continental shelf: the flat part of the sea floor next to land areas. Continental shelves, or large parts of them, may be exposed at times of low sea level.

Coombe deposits: the name given to *head deposits* in the *chalk* hills of southern England, including the North and South Downs and the Chilterns, and comprising lumps of chalk and *flint* set in a chalk mud, sometimes cemented into a hard mass called coombe rock.

Coprolite: fossilised human or animal excreta, usually surviving only in exceptional circumstances.

Cordate: literally 'heart-shaped', used in stone-tool typology to describe a form of *handaxe* with the maximum width in the lower third of the handaxe and convex sides. Narrower than the ovate handaxe.

Core: in stone-tool making, the nodule or cobble of stone material (such as *flint* or *quartzite*) from which *flakes* are removed. The core preserves the outlines of flake removals and may also show marked hollows (*negative bulbs of percussion*) which correspond to the *bulbs of percussion* seen on flakes.

Cortex: in stone-tool typology, the outer 'skin' sometimes present on a stone nodule or cobble. On *flint* the cortex tends to be chalky, while on *quartzites* and *Greensand chert* it is a more granular version of the internal rock.

Cortication: the formation of *cortex* on a stone nodule or cobble.

Cosmogenic nuclide dating: a relatively new *absolute dating* technique used in geology. Its application to the *Quaternary period* is not well established although the potential has been recognised. Cosmogenic nuclides are types of atoms formed in minerals at or near the Earth's surface due to cosmic radiation. The accumulation of cosmogenic nuclides provides a measure of the length of time that a surface has been exposed – somewhat similar to the way that the redness of someone's skin provides a measure of the length to time they have been exposed to the Sun. One application is the dating of *glacial moraines*. Assuming that the boulders in a moraine were shielded from exposure to cosmic radiation prior to deposition by the glacier but have been exposed since, the accumulation of cosmogenic nuclides in these rocks provides a measure of the time that has lapsed since deposition of the moraine.

Coversand: a wind-blown deposit, usually formed in *periglacial* conditions and mainly consisting of sand-sized particles. Similar to *loess* but more coarse grained.

Cranium: the bones of the skull that encase the brain, i.e. excluding the lower jaw (*mandible*).

Creep: in *geomorphology/sedimentology*, the very slow downslope movement of material.

Crested blade: in stone-tool technology, the first removal from a *blade core* that has been *crested* by the use of *alternate flaking* during core preparation.

Cresting: in stone-tool technology, the preparation of *facets* on a *blade core* by means of *alternate flaking* in order to ensure a predictable removal of the blade.

Creswellian: the name given to the major British *Late Upper Palaeolithic* stone-tool *industry*. The tools are typically made on *blades* (*Mode 4*), a characteristic type being the *Cheddar point*. The industry is a late regional variant of the *Magdalenian*, a cultural *tradition* widespread over much of Europe. Dating from the first half of the *Late Glacial Interstadial* (*Bølling Interstadial*), about 15,400–14,000 BP. The Creswellian is named after finds made in the caves at the *type site* of Creswell Crags, on the Derbyshire–Nottinghamshire border.

Cretaceous: the geological *period* lasting from 145 to 65 million years ago. The name comes from the Latin for 'chalk', and the Chalk of eastern and southeastern England – consisting of the *calcite* plates of countless millions of microscopic marine algae – symbolises the period. The primary importance of the Chalk for the much later *Ice Age* is that *flint* nodules formed within it and were exposed at outcrops, providing an excellent raw material for *Palaeolithic* knappers.

Cro-Magnon: a name given to the populations of <u>Homo sapiens</u> that occupied Europe during the *Upper Palaeolithic* ('Cro-Magnon Man'). Named after three human skulls and parts of skeletons found in 1868 at the Cro-Magnon rock shelter in France.

Cromer Forest-bed Formation (CF-bF): a series of deposits exposed discontinuously in cliffs along an 80 km length of the coasts of Norfolk and Suffolk. The deposits comprise sands and *organic* muds laid down by ancient rivers, including the *Bytham River*, draining midland and eastern Britain before the *Anglian glaciation* of around 450,000 BP. The cliffs are being actively eroded by the sea, and for more than 150 years impressive collections of the bones of extinct large *mammals*, together with the remains of beetles, *molluscs*, fruits, seeds and trees (from which the formation takes its name) have been made. It is now believed that the CF-bF represents at least six distinct temperate phases, spanning about 780,000–450,000 BP, interspersed by cold or *glacial* phases, possibly including the *Happisburgh Glaciation* (MIS 16). In 2005 the discovery of undoubted flint artefacts – simple *Mode 1 flake* tools – from CF-bF deposits at Pakefield, Suffolk, was announced. Multiple lines of evidence date these tools to around 700,000 BP, and they represent the earliest unequivocal evidence for human occupation north of the Alps. Environmental evidence shows that this occupation took place during a warm *interglacial* with a Mediterranean-type climate (including hippos). This interglacial (*MIS* 17?) has been termed the *Pakefield Interglacial*.

Cromerian: this term, which refers to an interglacial phase in the British *Pleistocene* sequence, has two usages. In its strictest sense, it refers to the interglacial deposits represented at West Runton in Norfolk (possibly *MIS* 15). This is sometimes described as the Cromerian <u>sensu stricto</u> (s.s.). Alternatively, the *Cromerian Complex* or Cromerian <u>sensu lato</u> (s.l.) refers to the whole of the four warm phases (MIS 19, 17, 15 and 13), and three intervening cold phases (MIS 18, 16 and 14), preceding the Anglian. This spans the long time period from about 787,000–474,000 BP. See also *Pakefield Interglacial* and *Happisburgh Glaciation*.

Cromer/Cromerian Complex: see *Cromerian*.

Cross-dating: a means of dating a deposit by means of comparing the contents (e.g. faunal remains or *artefacts*) of one deposit with another. The assumption is that deposits with similar contents – or containing the same *index fossils* or *indicator species* – will be of approximately the same date. Cross-dating can be used as a *relative dating* technique, to establish that two or more different deposits are of similar date, or to provide an approximate *absolute* date by comparing an undated deposit with one that has been dated by absolute methods.

Cyrogenic: relating to ice or the action of ice.

Cryology: the scientific study of snow and ice.

Cryosphere: the Earth's covering of ice.

Cryoturbation: the breaking, churning and mixing of the *active layer* in *permafrost* regions due to repeated freezing and thawing.

Culture: in *archaeology* the term 'culture' has a distinctive meaning. It is used to describe collectively archaeological *assemblages,* features (including structures and burials) and sites that are similar in character and occur over a limited geographical area within a specific time frame. It was originally believed that archaeological cultures belonged to, and were indicative of, distinct social groups such as 'peoples' or 'tribes'. Today it is realised that this is often not the case and that the relationship between archaeological cultures and meaningful social groups is, at best, problematic. Because of such problems the term is not as widely used today as in the past. A possible example of a culture from Palaeolithic Britain is the *Creswellian* culture, although this may be better considered an *industry* or a *tradition*.

Cut marks: in *archaeology*, the marks left on bones resulting from the use of stone tools. Commonly used to identify butchery. Examination of cut marks under a microscope can reveal the tell-tale profile of grooves made with a stone tool, enabling them to be distinguished from the gnaw marks of animals or scratches caused by other processes.

Dansgaard-Oeschger events (D-O events): periods of very rapid warming (short *interstadials*) during the *Last Glacial*, recorded in Greenland *ice cores*. Up to 25 D-O events took place between about 100,000 and 15,000 *BP*. Each event typically starts with an abrupt warming in Greenland of about 5 to 10°C in a few decades, followed by a gradual cooling over several hundred years and then a sharp return to the preceding cold conditions. The cause of D-O events is much debated but not known for certain. Named after the two scientists who first identified these events. Compare *Heinrich events*.

Débitage: in describing stone-tool technology, this term has two usages. Most commonly it is used to refer to the debris resulting from stone-tool manufacture, i.e. all the *flakes* and little chips not intended for use. Strictly speaking, this French term should be used to describe the process of *knapping* to remove usable pieces. Compare *façonnage*.

Deciduous: plants that lose their leaves in autumn.

Deep sea cores: cores drilled through the soft *sediments* that have accumulated on ocean floors. These cores are used for reconstructing past temperature and environmental change, for example by means of *oxygen isotope analysis*. See also *marine isotope record*. Compare *ice cores*.

Deflation: with reference to rocks and *sediments*, wind erosion.

Dendrochronology: tree-ring dating. The technique makes use of the fact that trees produce annual growth rings, which can be counted, and that the rings also form a pattern of wide and narrow rings, produced in years respectively favourable and unfavourable for growth. The patterning in ring formation allows wood from living and dead trees and timber to be matched, producing series that can now be taken back to around 12,000 *BP*. An important use of dendrochronology is to *calibrate* the *radiocarbon* curve.

Denticulate: in stone-tool technology, a *flake* or *blade retouched* to produce a saw-tooth edge.

Dentition: a term which describes the arrangement of teeth in an animal species. The study of dentition is important for several reasons: the teeth are the hardest part of the skeleton and therefore the most likely to survive; the type and arrangement of teeth is a major factor in identifying animal remains to species; the dentition of an animal reflects its diet; patterns of tooth eruption and the amount of wear on teeth can be used to estimate the age at death of an animal.

Devensian: the Devensian Glaciation or stage is the name given in Britain to the last major cold period of the *Pleistocene*, lasting from about 115,000–11,400 *BP*. It was not one long cold period but a period of generally cold but fluctuating climate encompassing several *Marine Isotope Stages* (*MIS* 5d to MIS 2 inclusive), including two major cold stages – MIS 4 and MIS 2 (*Last Glacial Maximum*) – with a generally warmer but very variable period in between (MIS 3). The Devensian is the equivalent of the *Weichselian* in Northern Europe, the *Würmian* in the Alps and the Wisconsinan in North America. The *Last Glacial* is a general term for this period.

Devonian: the geological *period* from about 416 to 359 million years ago.

Diagenesis: post-depositional physical and chemical changes in a *sediment*.

Diamicton: an unsorted sedimentary unit containing a mixture of sediments ranging in size from silt to boulders.

Diatoms: unicellular algae that live in ponds, lakes, estuaries and the sea. The cell wall is made of silica, which enhances the chances of survival of remains. Sensitive, amongst other things, to water temperature and salinity, they are used as *proxies* for regional climate and environmental change.

Diluvium: an old-fashioned term that used to be used for *Pleistocene* glacial and *fluvioglacial* deposits. It was coined when it was widely believed that these deposits resulted from Noah's Flood. Compare *alluvium*.

Dimlington Stadial: a name given in Britain to the major cold phase of the *Devensian,* corresponding with the first part of *MIS* 2 and also called the *Last Glacial Stadial*. The Dimlington Stadial lasted from about 27,000 to 15,400 *BP,* and includes within it the *Last Glacial Maximum*. It is followed by the *Windermere Interstadial*. Named after the *type locality* at Dimlington in Yorkshire.

Direct percussion: in stone tool making, the removal of *flakes* through directly striking the *core* with a *hard* or *soft hammer*. Compare *indirect percussion*.

Disarticulated: with reference to *fossil* bones, this term refers to the situation where the bones of the skeleton have become separated from one another and are found jumbled or isolated in the deposit that preserves them. Opposite of *articulated*.

Disc cores: in stone-tool typology, a *core* that has been rotated in the hand during several successive *flake* removals around its edge, resulting in a distinctive discoid shape. The flakes generated from this process will generally be broad in relation to their length.

Distal: far. The distal end of a long bone is that furthest from the body. The distal end of a *flake* is the end opposing that which bears the *striking platform*. The opposite of *proximal*.

Doggerland: the submerged landmass in the southern North Sea that was exposed during various periods of low sea level during the *Pleistocene*. At times this formed a vast plain connecting the present-day east coast of Britain with the northwest coast of continental Europe. Named after the Dogger Bank, which would have formed the 'Dogger Hills' at times of low sea levels.

Dorsal: in stone-tool making, the 'back' or 'upper' surface of a *flake* (or *blade*). This corresponds to the outer surface of the *core* from which the flake was struck, and may exhibit *dorsal flake scars* from flakes removed earlier in the *knapping* sequence. Compare *ventral*.

Dorsal flake scars: in stone-tool making, these are the evidence of previous flaking that may be seen on the back of a detached *flake*. These flake scars will show ripples indicating the directions previous flakes were removed in, and may also preserve negative *bulbs of percussion*. Study of dorsal flake scars helps to reconstruct how *cores* were flaked.

Downcutting: in the formation of *river terraces*, 'downcutting' refers to those phases when the river cuts down into the river valley and through deposits laid down at earlier stages.

Drift: in geology, 'drift' is a blanket term that covers any accumulation of glacial or *fluvioglacial* deposits. This is an old-fashioned term that was coined at a time when it was widely believed that these deposits resulted from material that had been carried by icebergs which drifted over flooded land and deposited their load as they melted. However, geological maps are still divided into those that show the 'drift' (or superficial) deposits and those that show the underlying 'solid' geology.

Drumlin: a streamlined, typically cigar-shaped ridge formed at the base of a *glacier* and orientated in the direction of former ice flow.

Dyke: in geology, a *bed*-shaped *igneous* or *clastic* intrusion which is discordant with (i.e. cuts though) the host beds. See also *sill*.

Early Middle Palaeolithic: in Britain, the *Middle Palaeolithic* record can be divided into two distinct parts, Early and Late, due to a long intervening period during which Britain was apparently deserted. The defining feature of the Middle Palaeolithic is the use of the *Levallois* technique of stone-tool manufacture. In Britain this first appears around 300,000 BP, at the transition from *MIS* 9 (the *Purfleet Interglacial*) to MIS 8, marking the beginning of the Early Middle Palaeolithic. The end of the period is marked by the beginning of the MIS 6 glacial about 186,000 BP, after which Britain was apparently abandoned until around 60,000 BP, the beginning of the *Late Middle Palaeolithic*. The Early Middle Palaeolithic is associated with the evolution of the *Neanderthals*; remains of early Neanderthals (unfortunately just teeth) have been found at Pontnewydd Cave, North Wales.

Early Pleistocene: an alternative term for *Lower Pleistocene*.

Early Upper Palaeolitihic: in Britain the *Upper Palaeolithic* record can be divided into two distinct parts, Early and Late, separated by the *Last Glacial Maximum*, during which Britain was deserted. The Early Upper Palaeolithic is associated with the first appearance of modern humans (Homo sapiens) in Britain and is tentatively dated from about 44,000 BP to around 30,000 BP. Three distinct stone tool industries have been recognised in this period – *leaf points, Aurignacian* and *Gravettian* – but the record is thin. The period is complicated by the possibility of overlap between *Neanderthals* and modern humans, with some authorities suggesting that the leaf points may have been made by Neanderthals. See also *Late Upper Palaeolithic*.

Ecofact: a blanket term sometimes used for the floral and faunal remains which survive on an archaeological site, in contrast to *artefacts*. Seeds, *pollen*, animal bone, insect and *mollusc* remains are all ecofacts.

Ecological succession: the succession of different plants and animals occupying an area, typically in response to climate change. For example as an *interglacial* period develops, arctic-alpine species will be replaced by a succession of different communities, perhaps culminating in a mixed-oak woodland *climax community*. The composition of each community will be governed not just by temperature but by the speed at which different species are able to colonise, competition and a host of other factors.

Ecology: the study of the interaction between organisms and their physical and chemical environment.

Ecosystem: an interdependent community of animals, plants and bacteria and their interrelated physical and chemical environment.

Ecotone: a transition zone between two different zones of plant and animal distributions or *biomes*, for example between *boreal* forest and *steppe* zones.

Eemian Interglacial: name given to the *Ipswichian Interglacial* or *Last Interglacial* in continental northwest Europe, about 127,000–115,000 BP.

Electron Spin Resonance (ESR) dating: a *radiometric* dating technique applied principally to the dating of tooth enamel. Newly formed tooth enamel does not contain any trapped electrons, but after a tooth is buried these begin to accumulate from natural radiation. Measuring the quantity of trapped electrons in a sample provides an estimate of how long it has been buried. The trapped electrons give rise to characteristic signals known as electron spin resonance, which is measured in the laboratory. The potential range of the technique is up to one million years, but there are various limiting factors and sources of potential inaccuracy.

Elsterian Glaciation: name given to the *MIS 12 glacial stage* in continental northwest Europe, about 474,000–427,000 *BP*. Equivalent to the *Anglian Glaciation* in Britain.

Eluvial horizon: a horizon in a soil from which mineral and organic matter has been lost by *leaching*. Compare *illuvial horizon*.

Encephalisation: brain expansion relative to body size.

End moraine: *moraine* formed at the end or snout of a *glacier*. Compare *terminal moraine* and *lateral moraine*.

Entrainment: the process by which material in the ground under a *glacier* becomes incorporated in the moving ice.

Eolith: Greek for 'dawn stone'. Used to describe very crudely chipped *cores* or *flakes* found in pre-*Pleistocene* or very early Pleistocene deposits. Once widely thought to be the oldest stone tools, they are now generally believed to have been chipped by natural agencies. However, the possibility that some are real *artefacts* may have to be reassessed in the light of the discovery of genuine simple flake tools at early sites such as Pakefield.

Eon: the longest unit of geological time, each one lasting hundreds of millions of years. Compare *era*, *period* and *epoch*.

Epigravettian: a cultural tradition following on from the *Gravettian* in parts of southern and eastern Europe, contemporaneously with the development of the *Solutrean* and *Magdalenian* in western Europe.

Epipalaeolithic: a term with a similar meaning to *Final Upper Palaeolithic* or *Terminal Upper Palaeolithic* and referring to the human societies and stone-tool industries at the very end of the Pleistocene and the beginning of the Holocene. The Epipalaeolithic merges with, and is sometimes used synonymously with, the *Mesolithic*.

Epiphyses: the articular ends of long bones or vertebrae. In adults these are fused to the shaft or main part of the bone but in juveniles they are separate bony masses. The degree of fusion of the epiphyses can be used to estimate the age at death of skeletal remains.

Epoch: the smallest major unit of geological time – the *Holocene* and *Pleistocene* are epochs within the *Quaternary Period*. Compare *eon, era* and *period*.

Era: an intermediate unit of geological time, being a sub-division of an *eon*. The present geological era, which began about 65 million years ago with the extinction of the dinosaurs, is called the *Cenozoic* (sometimes 'Cainozoic'), meaning 'recent life'. See also *eon, period* and *epoch*.

Erratic: a glacial erratic is a rock, generally larger than gravel size, that has been transported from its place of origin by glacial action (or sometimes by icebergs). The presence of erratics helps to determine the former extent and direction of movement of *glaciers*. Erratics of fine-grained rock were sometimes used as *toolstone* by *hominins*

Esker: a sinuous ridge of sand and gravel deposited by meltwater that flowed through channels beneath and inside a *glacier*.

ESR: see *Electron Spin Resonance dating*.

Eustasy: a uniform world-wide change in sea level, largely resulting from changes in the amount of water taken up by *ice sheets* and *glaciers*.

Eustatic rise: rise in sea level as a result of glacial melting and the release of the meltwater into the oceans. Contrast with *isostatic uplift*.

Exoskeleton: the hard outer covering of *invertebrate* organisms such as insects.

Fabric analysis: in geology, analysis of the orientation of *clasts* in a deposit.

Facets: in stone-tool technology, the marks left on a *core* or tool after a *flake* has been removed. Also known as *flake scars*.

Facies: a body of sediment that is characterised by a distinct combination of *lithological*, physical and biological properties, reflecting the process(es) of formation.

Facies log: a record of a sedimentary section.

Façonnage: in stone-tool making, the deliberate shaping of a stone tool by the removal of *flakes*. *Handaxes* were made by façonnage. See also *reduction*. Compare *débitage*.

Family: the major *taxonomic* rank in the *Linnaean system* or hierarchy that is below *order* and above *tribe*. For example humans belong to the *hominin tribe* within the *hominid* family within the *primate order*.

Fauna: a collective term for animal life. Also used to describe the community of animals occupying a given area or the *fossil* remains of these.

Federmesser: 'penknife' in German. Federmesser point (or Federmesser/Azilian point) is an alternative name for the *Final Upper Palaeolithic penknife points* found in Britain and dating roughly to the last part of the *Lateglacial Interstadial,* about 14,000–12,900 BP. The use of the German term draws attention to the close similarities between the German and British examples, and is a reminder that at this time the southern North Sea basin was not flooded and these two regions were connected by the vast plains of *Doggerland*.

Femur (plural **femora**): thigh bone.

Fennoscandian Ice Sheet: the name given to the principal north European *ice sheet* centred on Finland and Scandinavia but much more extensive at times of glacial maxima.

Fibula (plural **fibulae**): the smaller of the two bones of the lower leg. Compare *tibia*.

Ficron: a type of *handaxe* with a long, well-made point and concave sides. Name given by the famous French prehistorian François Bordes.

Final Upper Palaeolithic (FUP): the penultimate stage of the *Upper Palaeolithic*, followed by the *Terminal Upper Palaeolithic*, and occupying the last part of the *Lateglacial Interstadial* (*Allerød Interstadial*), from around 14,000–12,900 BP. In this period the *Creswellian* of the *Late Upper Palaeolithic* gives way to Final Upper Palaeolithic stone-tool *assemblages*, including *penknife point* (*Federmesser*) assemblages.

Firn: German for 'old snow'. Densely packed old snow, in which the crystals are partially fused together. An intermediate stage in the transformation of snow into ice.

Fissures: in stone-tool technology, the stress marks that radiate away longitudinally from the *point of percussion*.

Flake: in stone-tool making, a flake is any piece struck from a *core* or another flake. Flakes are the simplest form of stone artefact, often preserving evidence of human manufacture such as a *bulb of percussion* and a *striking platform*. Flakes (especially flint flakes) are very sharp and could have been used for a variety of cutting tasks.

Flake scar: in stone-tool making, this is the scar left by the removal of a flake from a *core* or another *flake*. The scar may exhibit distinctive features such as a *negative bulb of percussion*. Also known as *negative flake scars* or *facets*.

Flaking angle: in stone-tool typology, a measure of the angle at which a *flake* is struck from a *core*. This reveals information about flaking technique, e.g. narrow flaking angles are typical of *soft hammer* techniques and wide flaking angles of *hard ham*mer techniques. The *internal flaking angle,* which is the angle between the flake's *butt* and its *ventral* surface, is that most usually measured.

Flandrian: an alternative name (not much used today) for the *Holocene epoch*, the present i*nterglacial* episode beginning about 11,400 years ago. The name derives from the *marine transgression* of the Flemish coastal plain associated with the present warm period. See also *postglacial*.

Flint: a hard cryptocrystalline silicate that occurs as nodules in sedimentary rocks such as *chalk* and *limestone*. It is commonly dark grey or black in colour and the nodules are covered in a chalky outer coating called *cortex*. Due to its structure (similar to glass) flint fractures in predicable ways and is excellent for making stone tools.

Flint knapping: see *knapping*.

Floodplain: the part of a river valley subject to periodic flooding. As the floodwaters recede their *sediment* is deposited as *alluvium*.

Flora: a collective term for plant life.

Flot: a term used to describe the material which floats on the surface of the water (or other liquid) during the *flotation* process.

Flotation: the process of extracting *organic* remains, such as the remains of plants, insects and *molluscs*, from a sediment by means of immersing the sediment in water (or another liquid medium) so that the remains (the *flot*) float to the surface. Flotation machines can be complex or simple (a bucket) but the basic principle remains the same. See also *wet sieving*.

Fluorine test: a technique for establishing the relative date of bones from the same deposit. Buried bones absorb fluorine from the ground water, with progressively more fluorine being absorbed the longer the bones have been buried. Chemical analysis can determine the fluorine content of bones, and bones that have been buried for different lengths of time will have different fluorine contents. It was fluorine testing that definitively established Piltdown Man as a forgery, the bones from the human skull being demonstrably much younger (i.e. having a much lower fluorine content) than the bones of early Ice Age *mammals* with which they were supposedly associated. Because the fluorine levels in bones depends on local conditions, the method cannot be used to establish *absolute* dates.

Fluvial: relating to rivers.

Fluvioglacial: the processes and landforms relating to glacial meltwater streams. Also *glaciofluvial*.

Foraminifera: single-celled marine animals which secrete a carbonate shell; 'forams' for short. The different proportions of oxygen isotopes in the shells of forams cored from the sea bed is used to establish the *marine isotope record*.

Formation: in *lithostratigraphy*, a stratigraphic unit comprising several *members* which accumulated sequentially during a major depositional event (e.g. a glacial episode or *marine transgression*).

Fossil: the buried remains of a plant or animal, usually preserved within a sediment. In *geology*, the term is sometimes restricted to remains that have become mineralised, but this does not generally apply to *Quaternary* remains. See also *microfossil, macrofossil, sub-fossil* and *trace fossil*.

Fossil beach: see *raised beach*.

Gastropod: snail. Gastropods are members of the slug and snail *class* Gastropoda, part of the *phylum* Mollusca.

Gelifluction: a term often used to describe *solifluction* in a *permafrost* zone. The spring thaw of the *active layer* above the permafrost causes the upper layer of the ground to become saturated with meltwater. On slopes this will cause downslope flow of the soil, together with larger stones and boulders.

Genus (plural **genera**): the rank in the *Linnaean system* of biological classification between *species* and *family*. A genus is a group of species that share a common ancestor and are more closely related to one another than to any other genus group within their family. Species are named by giving the genus name first and the species name second, e.g. Homo (genus) sapiens (species).

Geochronology: the scientific study of the ages at which geological deposits were laid down and when geological events took place.

Geology: the broad science that deals with the study of the Earth. Compare *geomorphology, archaeology, palaeontology* and *Quaternary science*.

Geomagnetic reversals: changes in the orientation of the Earth's magnetic field such that the positions of magnetic north and south become reversed. When the orientation of the field is the same as today (i.e. magnetic compasses point north) this is termed 'normal' polarity and when the orientation flips this is called 'reversed' polarity. Each change in polarity takes thousands of years to complete. The cause of the reversals is not known for certain but is thought to be related to the chaotic motions of liquid metal at the Earth's core. Each major period of polarity in one or the other orientation is called a 'Chron'. The present period of normal polarity is called the *Brunhes Chron* and the preceding period of mainly reversed polarity is called the *Matuyama Chron*. As iron minerals in molten lava orientate themselves on the magnetic poles but their orientation becomes 'frozen' when the rock solidifies, in combination with absolute dating methods, cross-correlations and stratigraphic inference, geomagnetic reversals can be used for dating purposes.

Geomagnetism: the Earth's magnetic field and related phenomena.

Geomorphology: the science that deals with physical features on the surface of the Earth (e.g. river valleys) and how they are formed.

Geotechnical: relating to the geological aspects of engineering.

Gipping Stage: a geological stage defined from deposits in East Anglia and correlated with the *Wolstonian* in the conventional British chronology for the *Quaternary*. It is now realised that the Wolstonian/Gipping 'stage' as originally conceived conflates terrestrial deposits probably representing three separate cold *marine isotope stages* (with two intervening warm *interglacials*) between about 364,000 *BP* and 127,000 BP, so neither term is now used.

Glacial: as a noun, a major cold stage in the *Quaternary* characterised by the growth and expansion of *ice sheets*.

Glacial drift: material transported by *glaciers* or glacial rivers.

Glacial erratic: see *erratic*.

Glacial maximum: the peak of a *glacial* period when temperatures are at their lowest and *ice sheets* at their maximum extent.

Glacial outwash: material (e.g. gravel) deposited by glacial meltwater streams at or beyond the edge of a *glacier*. Due to their origin, glacial outwash deposits will very seldom contain significant fossil or archaeological material.

Glacial rebound: see *isostatic uplift*.

Glacial till: see *till*.

Glacier: a mass of ice that deforms under its own weight and flows downslope.

Glaciofluvial: the processes and landforms relating to glacial meltwater. Also *fluvioglacial*.

Glaciology: the study of the physics of glaciers.

Gorge: with respect to *artefacts*, a small double-pointed sliver of bone believed to have been used for catching fish. The gorge is tied in the middle with a line so that it will lodge in the fish's mouth when the line is pulled.

Gravettian: a European *Upper Palaeolithic industry* dating from around 33,000 *BP* to 25,000 BP. Characterised by projectile technology and elaborate burials (including the 'Red Lady' of Paviland in Wales). Follows the *Aurignacian* and ends during the severe conditions of the *Last Glacial Maximum*, where it is succeeded in Europe west of the Rhône by the *Solutrean* and the *Magdalenian*, but in southern and eastern Europe by the *Epigravettian*.

Grazer: a *herbivore* whose diet consists mainly of grass.

Great Ice Age: a name sometimes given to the *Quaternary* Ice Age (i.e. the *Pleistocene*), as opposed to ice ages in earlier geological *periods*, some of which, however, were more severe.

Great Interglacial: a name sometimes given to the *Hoxnian Interglacial*.

Greensand chert: a fine-grained cryptocrystalline sedimentary rock, similar to *flint* though of more variable appearance and texture. Greensand *chert* is also suitable for stone-tool manufacture, and Palaeolithic assemblages of this material are known mainly from the southwest of England.

Groove-and-splinter: a technique used in the *Upper Palaeolithic* for extracting tool blanks from an antler. A *burin* is used to incise two parallel grooves in the beam of an antler, penetrating through the tough outer layer into the softer layer beneath. The splinter is then prised out and may be further worked into a range of tool types, for example a *barbed point*.

Günz Glacial: the name given to an *Early Pleistocene glacial* identified in the Alps. For much of the 20th century the Alpine sequence of glaciations – Günz-Mindel-Riss-Würm – formed a framework for Europe as a whole, but this has now been recognised as a gross oversimplification and the practice has been dropped, although it will still be found in older literature.

Günz-Mindel Interglacial: the interglacial between the Günz and Mindel glacials, as recognised in the Alps. The rough equivalent in Britain is the *Cromerian Complex*.

Haft: the handle or shaft to which some stone tools were attached. Hafting is the process of attaching, say, a stone spear point to a wooden shaft. Evidence for hafting does not appear before the *Middle Palaeolithic*.

Half-life: in *radiometric dating*, the time it takes for a radioactive *isotope* to disintegrate one half of its radioactive atoms. The length of the half-life of a radioactive isotope (e.g. *radiocarbon*) is important in determining the chronological range of a particular radiometric dating technique.

Hammer: in stone-tool making, a hammer is any object used to strike *flakes* from a *core* or another *flake*. A hammer is therefore a tool used to make a tool. See also *hard hammer* and *soft hammer*.

Handaxe: a *Palaeolithic* stone tool flaked on both faces to produce a long and durable cutting edge. Handaxes made during the *Lower Palaeolithic* are known as *Acheulean* handaxes, while certain specific forms of handaxe continued to be made in the *Middle Palaeolithic* and are known as *Mousterian* or *bout coupé* handaxes. Handaxes and related tool types are also known as *bifaces* and define *Mode 2 lithic* technology. Handaxes are the most commonly recovered type of stone tool found in England as they are large and clearly humanly made, and therefore relatively easy to spot.

Happisburgh Glaciation: a recently postulated glaciation equated with *MIS* 16 (about 659,000–621,000 *BP*). The principal evidence for this is the Happisburgh *Till*, which appears in deposits of the *Cromer Forest-bed Formation* exposed along the Norfolk/Suffolk coast. An alternative interpretation is that the Happisburgh Till represents one of the pulses of ice advance of the *Anglian Glaciation* (MIS 12). Named after the village of Happisburgh (pronounced Hazeborough) on the Norfolk coast.

Hard hammer: in stone-tool making, a cobble used to strike *flakes* from a *core* in *direct percussion* technique. *Quartzite* cobbles make very good hammers for *flint knapping*. Compare *soft hammer*.

Head deposits: in valleys which have experienced *periglacial* conditions the process of *gelifluction* can lead to the build up of deep deposits of material, including frost-shattered stone, brought down from the surrounding slopes.

These are termed head deposits and usually have their long axes orientated downslope, reflecting the flow direction. See also *coombe deposits*.

Heinrich events: episodes of extreme cold during the *Last Glacial* when the circulation of the North Atlantic was altered by the influx of armadas of icebergs associated with the collapse of part of the *Laurentide ice sheet*. Six Heinrich events have been recognised, the earliest (H6) occurring about 60,000 *BP* and the most recent (H1) about 16,800 years ago (although the *Younger Dryas Stadial* may also be a Heinrich event). The onset of Heinrich events was very rapid and they each lasted around 750 years. Some but not all cold Heinrich events preceded warm *Dansgaard-Oeschger events*. Heinrich events are named after the marine geologist Helmut Heinrich who first described them from evidence of rocks dropped by the icebergs in the North Atlantic.

Herbivore: an animal that eats predominantly or exclusively plant matter.

Holocene: the geological *epoch* corresponding to present *interglacial*, which began around 11,400 years ago. Together the Holocene and *Pleistocene* epochs make up the *Quaternary* geological *period*. Equivalent terms are *Flandrian* and *Postglacial*.

Holotype: in *palaeontology*, the specimen used as the basis of the original published description of a *taxonomic* group and later designated as the *type fossil* for that group.

Holsteinian Interglacial: the name given to the *Hoxnian Interglacial* (*MIS* 11, about 427,000–364,000 *BP*) in continental northwest Europe.

Homeostasis: the maintenance of stable equilibrium.

Hominid (Hominidae): in biological *taxonomy*, traditionally describes all *bipedal* apes ancestral to and including modern humans, as well as many more now extinct species of bipedal apes not directly on this line but closely related to it. In the light of recent genetic studies, which show our close relationship to other apes, the *taxon* 'hominid' now includes chimpanzees, bonobos, and gorillas. Most literature before about 2000 uses 'hominid' in the traditional sense. Compare *hominin*.

Hominin (Hominini): in biological *taxonomy*, the term now used to describe modern humans and their immediate *bipedal* fossil ancestors and relatives (*hominids* in the traditional sense). The hominin line split from the line leading to modern chimpanzees about 5–7 million years ago, according to genetic data. In the *Linnaean hierarchy*, hominid describes the *family* and hominin describes one of the *tribes* within that family. Within the hominin tribe there are several *genera*, including Ardipithecus, Australopithecus, Paranthropus and Homo. All the species within all these genera are now extinct, with the exception of Homo sapiens.

Homo: the human *genus*. It is not agreed how many *species* should be included in the genus Homo but amongst the least controversial are Homo erectus, Homo neanderthalensis and Homo sapiens.

Homo antecessor: 'Pioneer Man'. A human species defined from fossil remains found at Gran Dolina, Sierra de Atapuerca, northern Spain, and dated to around 800,000 *BP*. The evolutionary relationships, and in particular whether the species really is distinct from Homo heidelbergensis, are much debated. Antecessor is associated only with *Mode* 1 (*core* and *flake*) technology and is a possible candidate for the maker of the flake tools found at Pakefield, Suffolk, and dated to around 700,000 BP.

Homo ergaster: 'Working Man'. A human species often considered to be an early, exclusively African form of Homo erectus from which Asian Homo erectus evolved. Dated between about 1.9 and 1.5 mya. Many scholars do not accept the species distinction between ergaster and erectus.

Homo erectus: 'Upright Man'. Considered by some to be the earliest species that can be confidently assigned to the genus Homo. Found from Africa (if merged with Homo ergaster) to Asia ('Java Man' and 'Peking Man') and possibly to Europe (depending on definition). A possible date range from 1.9 *mya* to 27 *kya*, although both the early and late dates are controversial. Associated with the first major expansion of humans outside Africa, advances in stone-tool technology, e.g. *handaxes* (but not in east Asia), and possibly the first controlled use of fire. Very similar, if more robust, to modern humans from the neck down, erectus is also believed to have had a similar life-history pattern, e.g. extended childhood. However, most scholars believe that erectus would not have possessed the ability for grammatical speech.

Homo georgicus: a putative human species defined from *fossil* remains found at Dmanisi, Republic of Georgia, in the Caucasus. The remains, excavated from the early 1990s onwards, comprise several skulls, jaws and parts of skeletons dated to around 1.7 *mya*. However, many scholars dispute the assignment of these fossils to a new species, and see them as *Homo ergaster*, early *Homo erectus* or even more primitive species of *Homo*. It is possible that more than one species is represented. The associated tools comprise simple *flakes*, *scrapers*, and *choppers* (*Mode 1*). Whatever the correct assignment of these fossils, their presence on the fringe of Europe at such an early date is of great significance.

Homo heidelbergensis: 'Man from Heidelberg'. An ancient human species that inhabited Europe and Africa between about 600,000 *BP* and 300,000 BP, although there are different interpretations of the age and distribution of this species. In one prominent interpretation of the *phylogeny*, Homo heidelbergensis evolved from *Homo erectus* around 600,000 BP, probably in Africa, and then spread to Europe. In Europe heidelbergensis eventually evolved into the Neanderthals while in Africa the species evolved into modern humans. Named after a lower jaw found in 1907 during quarrying at the Mauer sand pit near Heidelberg. Homo heidelbergensis may have been responsible for introducing the manufacture of *handaxes* into Europe. The shin bone and two teeth recently found in a quarry at Boxgrove in West Sussex are attributed to Homo heidelbergensis. These human remains are dated to around 500,000 BP and are associated with numerous finely-made flint handaxes.

Homo neanderthalensis: 'Neanderthal Man' ('Man from the Neander valley [thal]'), an ancient human species that inhabited Europe and adjacent parts of Asia from around 300,000 *BP* to about 30,000 BP. The Neanderthals are thought to have evolved in Europe from populations of *Homo heidelbergensis* and therefore the distinction between late heidelbergensis and early Neanderthal is not easy to draw. The classic Neanderthals, including the type specimen found during quarrying in the Neander valley in 1856, date from the period 70,000–30,000 BP. Physically adapted to living in cold climates, the robust skeleton of the Neanderthals reflects a tough, physically demanding lifestyle. The size of the Neanderthal brain was as big as that of modern humans, but it is debated whether they were capable of grammatical speech and their material culture was simpler than that of modern humans, with figurative art objects lacking. In Europe, the Neanderthals are associated with *Middle Palaeolithic* stone-tool techniques and industries, including the *Levallois* technique and the *Mousterian*.

Homo sapiens: 'Wise Man', modern humans. It is now generally believed, on the basis of both fossil and genetic evidence, that Homo sapiens evolved in Africa from *Homo heidelbergensis* over the period 300,000–130,000 *BP*. Thus by about 130,000 years our own species had evolved but was largely confined to Africa. Then, around 60,000 BP Homo sapiens expanded out of Africa into Asia and on into Australia (which was colonised for the first time), and also (from about 45,000 BP) into Europe. In Asia and Europe Homo sapiens seems to have replaced the indigenous human populations, late *Homo erectus* and *Homo neanderthalensis* respectively. There is no good reason to believe that the modern humans that lived around 60,000 BP differed in any important sense from ourselves either biologically or in terms of their mental capacity. In Europe the arrival of modern humans is associated with the appearance of *Upper Palaeolithic* technology (including figurative art). The modern humans that occupied Europe during the Upper Palaeolithic are sometimes called the *Cro-Magnons*.

Horizon: in geology or archaeology, a layer of *sediment* representing a particular interval of past time.

Hoxnian Interglacial: the name given in Britain to a major *interglacial*, following the *Anglian Glaciation*, and now correlated with *MIS* 11, about 427,000–364,000 *BP*. This interglacial is rich in *Lower Palaeolithic* sites of both the *Acheulean* (*handaxe*) and *Clactonian* (non-handaxe) *industries,* although these are very largely confined to the south and east of the country. The Swanscombe partial human skull, discovered in a quarry in Kent, belongs to this interglacial and is transitional in some respects between *Homo heidelbergensis* and the Neanderthals. An alternative name for the MIS 11 interglacial in Britain is the *Swanscombe interglacial*. This alternative name is sometimes used because the Hoxnian interglacial was originally defined on the basis of *pollen biostratigraphy*, including at the *type site* of Hoxne in Suffolk (a brickearth quarry), and it is now realised that some deposits called 'Hoxnian' in fact probably date to later interglacials. The Hoxnian is also sometimes called the *Great Interglacial*, and in continental northwest Europe the *Holsteinian*.

Humerus (plural **humeri**): upper bone of the forelimb or arm.

Ice Age: In this book 'Ice Age' is used as an informal synonym for the *Pleistocene epoch*, encompassing both *glacial* and *interglacial* intervals. However, other authors sometimes use the term (especially in lower case – 'ice age') as an informal synonym for a *glacial* interval within the Pleistocene. Furthermore, there were several 'Ice Ages' during the course of Earth history, and where there is possibility of confusion about the Ice Age in question, then the term should be qualified by giving the relevant geological period, e.g. 'Ordovician/Silurian Ice Age', 'Pleistocene Ice Age', etc..

Ice cap: a *glacier* similar to an *ice sheet* but smaller (less than 50,000 square kilometres and sometimes only a few square kilometres in extent)

Ice cores: cores drilled into *ice sheets* and *glaciers* for the purpose of palaeoclimatic and palaeoenvironmental reconstruction. The annual snow fall forms layers within the ice and *oxygen isotope analysis* of the layers provides a *proxy* for past temperature changes. The snow fall each year also traps bubbles of atmospheric gas, ash, wind-blown dust and other inclusions, providing proxies for atmospheric composition, volcanic eruptions, desert extent and forest fires amongst other things. The most important cores are those that have been drilled in Greenland, where they are up to 3000 m long and cover a period back to 123,000 *BP* (the time of the *Last Interglacial*) and Antarctica, where recent cores have provided a record back to 720,000 BP.

Ice sheet: a continental-sized *glacier*, covering at least 50,000 square kilometres, that is dome shaped with a flow of ice outward from the centre. There are only two large ice sheets today, over Greenland and Antarctica, but there were a number during the Ice Age, the most significant of which were the *Fennoscandian* over northern Europe and the *Laurentide* over much of North America.

Ice wedge: in *periglacial* areas, seasonal temperature changes can cause cracking of the ground. Cracks that are filled up with ice are called 'ice wedges'. If the climate then warms up, the ice melts and the wedges fill up with sediment to form 'ice wedge casts'. Such casts, which are found in many areas of England, are testimony to the former presence of *permafrost*.

Igneous: Igneous rocks are rocks of volcanic origin, formed by the crystallization of once molten magma or lava. Examples include granite, dolerite, *andesite, basalt* and various *tuffs*. Fine-grained igneous rocks such as andesite and basalt are suitable for making stone tools, although not as good for this purpose as *flint*.

Illium: the largest bone of the pelvis, forming the upper and side parts of the pelvic girdle surrounding the birth canal.

Iluvial horizon: a layer in a soil where minerals and organic materials have accumulated due to *leaching*. Compare *eluvial horizon*.

Imbricated clast: a *clast* in a river bed or *till*, with the flat surface of the clast dipping in a roof tile-like manner.

Incisor: a narrow-edged tooth at the front of the mouth, adapted for biting in most *mammals*. In elephants (including mammoths) the upper incisors have become tusks.

Index fossil: a *fossil* characteristic of, and restricted to, an assemblage zone.

Indicator species: a biological *species* that provides an indication of some trait of the environment in which it is found. For example, some species are indicators of climatic conditions, while certain species of vole are indicators of the date range of the sediments in which they are found. See also *Vole Clock*.

Indirect percussion: in stone-tool making, the use of a *punch* (usually of antler or bone), hit with a *hammer*, to remove *flakes* or *blades* from a *core*. This allows greater precision and control in flaking than *direct percussion*.

Industry: with reference to stone tools, an industry is a distinctive configuration of artefact types and type frequencies that recurs amongst two or more *assemblages*. In practical terms an industry is defined by the finding of a substantial number of similar stone tool assemblages at different places.

Insect analysis: the use of insect remains (especially beetles) from ancient deposits to reconstruct past climates and environments. Compare *pollen analysis* and *molluscan analysis*.

***In situ*:** in place.

Insolation: the solar radiation received at any particular area of the Earth's surface, varying with weather, latitude and periodic variations in the Earth's orbit. From INcoming SOLar radiATION.

Interglacial: a major warm phase within the *Quaternary* characterised by the melting of ice sheets, rise in sea levels and an expansion of plants and animals adapted to warmer conditions. Compare *glacial*.

Internal flaking angle: see *flaking angle*.

Interstadial: a relatively warm interval within a *glacial* episode, but not sufficiently long or pronounced to be classed as an *interglacial*. Compare *stadial*.

Invertebrate: an organism without a backbone (e.g. insects).

Involution: contortion of surface material, such as soil, by processes such as creep, slumping, loading and expulsion of water – all common in *periglacial* regions.

Ipswichian: the name given in Britain to the last *interglacial* before the present one, correlated with *MIS* 5e and dating to about 127,000–115,000 *BP*. Summer temperatures were about four degrees Celsius higher than today, sea levels were high and Britain was an island with hippopotamus amongst the warm-climate fauna to be found, but it appears that there was no human occupation. Alternative names are the *Last Interglacial* and, in northwest continental Europe, the *Eemian*.

Isostasy: the process whereby the Earth's crust floats on an underlying, more fluid, mantle in a state of near equilibrium. However, the weight of vast *ice sheets* depresses the crust into the mantle and when the weight is released with the melting of the ice the land slowly returns to equilibrium, a process known as *isostatic uplift* or *glacial rebound*. This process affects relative sea levels and the formation of *river terraces*. Compare *eustasy*.

Isostatic uplift: the rise of landmasses resulting from the removal of the weight of *ice sheets*, also known as *glacial rebound*.

Isotopes: atoms of the same chemical element but with different atomic weights. The different isotopes of the same element have the same number of protons but a different number of neutrons, those with more neutrons having a greater atomic mass. Isotopes are labeled by writing the atomic mass in front of the symbol for the element – thus ^{18}O stands for the oxygen isotope with an atomic mass of 18 in contrast to the lighter ^{16}O. The different proportions of these two isotopes in sea water and glacial ice is used in *oxygen isotope analysis* for the reconstruction of past climates. The decay of radioactive isotopes into different isotopes or elements is the basis of various kinds of *radiometric dating*.

Ivory: the name given to the *tusks* of elephants (including mammoths) and some other *mammals* when these are used by humans as a resource, e.g. as a material used for carving.

Jerzmanovice leaf points: name given to a type of leaf-shaped point made on flint blades, dating to the very beginning (pre-*Aurignacian*) of the British *Early Upper Palaeolithic*, c. 44,000–40,000 *BP*. The name comes from a cave site in Poland, and as the name suggests this type of leaf blade-point is found in continental Europe as well as Britain. They were probably used as spear points. There is some debate as to whether they were made by *Homo sapiens* or late Neanderthals.

Jurassic: the geological *period* from about 200–145 million years ago.

Kame: a steep-sided hill of sand and gravel of glacial origin. A kame is formed when a deposit of material collects in a depression or crevasse in the surface of a *glacier*; as the stagnant ice melts away, the deposit is lowered to the valley floor forming a mound.

GLOSSARY: AN ICE AGE DICTIONARY

K/Ar dating: see *potassium-argon dating*.

Kettle hole: a bowl-shaped depression within an area covered by *fluvioglacial* deposits, often containing a pond. A kettle hold forms when a piece of ice buried beneath fluvioglacial material melts, causing the layer of debris above where the ice was to slump into the hole created.

Kingdom: in biological classification, the highest rank in the traditional *Linnaean system,* above *phylum*. For example humans (and innumerable other *species*) belong to the chordate phylum of the animal kingdom.

Knapping: the process of making stone tools by chipping and flaking. A maker of stone tools is called a knapper.

Kubiena tins*:* see *monolith tins*.

Kya: abbreviation for 'thousand years ago', e.g. 80 kya is 80,000 years ago.

Kyr: abbreviation for 'thousand years', e.g. 'the glacial lasted 80 kyr'.

Lacustrine: relating to lakes.

Lag deposit: in geology, a residue of coarse *clasts* from which finer clasts have been removed by wind or water action.

Lamina (plural **laminae**): in geology, a layer that is no more than 1 cm thick.

Last Glacial: a general name for the most recent glacial period, about 115,000–11,400 *BP,* corresponding to the *Devensian* in Britain, the *Weichselian* in Northern Europe, the *Würmian* in the Alps and the Wisconsinan in North America.

Last Glacial Stadial: an alternative, more general name for the *Dimlington Stadial*.

Last Glacial Maximum (LGM): the maximum advance of the *ice sheets*, between about 27,000–16,000 years ago, during the most recent (*Devensian*) *glacial*. Britain was deserted by humans during this very cold period, with much of the country covered by ice sheets.

Last Interglacial: alternative name for the *Ipswichian* or *Eemian interglacial*, about 127,000–115,000 *BP*.

Lateglacial: name given to the last stage of the *Devensian Glacial*, from about 15,400 years ago until the beginning of the *Holocene* about 11,400 years ago. The Lateglacial corresponds to the *Windermere Interstadial* and the subsequent *Loch Lomand Stadial* combined.

Lateglacial Interstadial: an alternative, more general name for what is called the *Windermere Interstadial* in Britain or the *Bølling-Allerød Interstadial* in continental northwest Europe, about 15,400-12,900 *BP*.

Late Middle Palaeolithic: in Britain, the *Middle Palaeolithic* record can be divided into two distinct parts, Early and Late, due to a long intervening period during which Britain was apparently deserted. The Late Middle Palaeolithic dates from about 60,000 *BP*, when Britain was reoccupied after a long period of human absence, to about 44,000 BP, the beginning of the *Upper Palaeolithic* (although with potential overlap). The humans occupying Britain during this period were classic Neanderthals (although their physical remains have not been found in this country) and most of the stone tools found can be grouped within the *Mousterian of the Achuelean Tradition (*MAT), the most distinctive element being *bout coupé handaxes*.

Lateral moraine: a *moraine* that forms along the side of a glacier. See also *marginal moraine*.

Late Upper Palaeolithic (LUP): in Britain the *Upper Palaeolithic* record can be divided into two distinct parts, Early and Late, separated by the *Last Glacial Maximum* (LGM), during which Britain was deserted. Following the LGM, reoccupation began perhaps around 15,400 years ago, marking the beginning of the Late Upper Palaeolithic. In terms of climate this relatively warm period is known at the *Lateglacial* or *Windermere Interstadial*. The most characteristic stone tool industry of the period is the *Creswellian*. The Late Upper Palaeolithic gives way to the *Final Upper Palaeolithic* perhaps around 14,000 *BP*.

Laurentide Ice Sheet: the main *ice sheet* covering North America during the Ice Age.

Leaching: in *geology*/soil science, the removal of soluble minerals from a soil, sediment or rock by percolating water.

Leaf points: see *Jerzmanovice leaf points*.

Lens: in geology, a thin stratigraphic layer tapering at the ends.

Lenticular: lens-shaped.

Levallois flake: a *flake* of predetermined shape produced by the *Levallois* technique.

Levallois point: triangular stone point produced by the *Levallois* technique. These may have been often used as spear points.

Levallois technique: in stone-tool making, a method of producing *flakes* and points of predetermined shape by detaching them from a carefully prepared *core*. The core is sometimes called a *tortoise core* on account of its shape. The introduction of this technique, about 300,000 *BP*, defines the beginning of the *Middle Palaeolithic* period in Europe, where the use of the technique is associated with the Neanderthals. The Levallois technique is an example of *Prepared Core Technology (PCT)*, which is also called *Mode 3* technology. Levallois is named after 19th-century discoveries at the *type site*, Levallois-Perret, a suburb of Paris.

Levee: a ridge-shaped bank of sediments along the side of a river channel, deposited when the water overflowed the sides of the channel.

Limestone: a *sedimentary* rock made of calcium carbonate and mainly laid down in marine conditions.

Linnaean system (or taxonomy): the method of classifying living things originally devised by the Swedish naturalist Carolus Linnaeus (1707-1778) and still used, with modifications, today. It has two main elements. The first is to classify all organisms according to a hierarchy of which the main *taxa*, proceeding in order from the most inclusive level to the most exclusive level, are (with the grey wolf as an example) *Kingdom* (Animalia), *Phylum* (Chordata), *Class* (Mammalia), *Order* (Carnivora), *Family* (Canidae), *Genus* (Canis) and *Species* (Canis lupus). Several other levels have since been inserted into the hierarchy, such as Superfamily and *Tribe*, but this does not affect the general principle. Although Linnaeus was unaware of it, the reason that this 'Russian doll' system works so well, as Charles Darwin and others realised, is because it reflects the process of 'descent with modification' by which species evolved. The second main element in the Linnaean system is the principle of *binominal nomenclature*, whereby each species is given a unique name (sometimes called its 'Latin name' or 'scientific name') comprising the genus name and a specific name. The genus name begins with a capital letter and the specific name with a small letter, and the whole name is usually given in italics (but here underlined because italics is used for cross-referencing) – for example, Canis lupus. In scientific papers the name of the person who first gave a valid description of the species may follow the species name, sometimes together with the date of the description – in the case of the grey wolf, Canis lupus Linnaeus,1758.

Lithic: relating to stone (from the Greek 'lithos'). Somewhat ungrammatically, archaeologists commonly refer to stone tools as 'lithics', and indeed there is a specialist journal devoted to all aspects of stone-tool technology called 'Lithics' (see Further reading: Journals).

Lithology: description of sediments and rocks on the basis of physical characteristics such as colour, structure and particle size.

Lithostratigraphy: the description and interpretation of stratigraphic sequences (e.g. as exposed in a quarry face) in terms of sediment properties (colour, particle size and shape, *sorting* etc.) and physical formation processes (deposition by rivers, *glaciers* etc.). The principle is to move from objective description to interpretation trying not to confuse the two. In a typical situation the basic stratigraphic units are *beds,* which are grouped into *members* and then into *formations*. See also *facies*.

Load: in connection with rivers, the material (silt, gravel etc.) carried by rivers.

Loam: a soil comprised of sand, silt and clay in roughly equal proportion.

Loch Lomond Stadial: the name given in Britain to the *Younger Dryas Stadial*, about 12,900–11,400 *BP,* the last major cold event of the *Devensian* glacial, following the *Windermere Interstadial* and immediately preceding the current *Holocene Interglacial.* The *stadial* takes its name from a large mountain icefield that developed in the western Highlands of Scotland at the time, an event that is called the Loch Lomond Readvance, and which was the last occasion when *glaciers* existed in the British Isles.

Lodgement till: debris (*till*) that is deposited beneath *glaciers* when the material being transported by the glacier becomes stuck or 'lodged' in the underlying bed. Lodgment till is often associated with the retreat phase of glaciers and the *clasts* tend to be relatively rounded, as a result of grinding at the ice/bed interface, compared with the clasts in *ablation till.*

Loess: a silty, unconsolidated wind-blown deposit. During the Pleistocene, large amounts of *outwash* material accumulated along the margins of the *ice sheets.* In the prevailing windy and dry conditions, huge quantities of fine-grained material was picked up, often carried great distances, and then deposited as loess. In England, loess deposits, rarely more than 2 m in depth, cover parts of East Anglia and the London basin, where they may become incorportated into *brickearth.*

Lower Palaeolithic: the name given to the earliest of the three subdivisions of the *Palaeolithic*, or Old Stone Age, in Europe. In Britain, on present evidence, the Lower Palaeolithic dates from about 700,000 years ago until about 300,000 years ago.

Lower Pleistocene: The earliest of the three subdivisions of the *Pleistocene* geological *epoch,* dated from 1.8 mya to 0.78 mya (also known as the *Early Pleistocene*). The transition to the *Middle Pleistocene* is marked by the *Brunhes-Matuyama geomagnetic reversal*. Although these dates have the authority of the International Commission on Stratigraphy, which aims to ensure consistency of usage amongst scientists, the dates are otherwise somewhat arbitrary (but see *Middle Pleistocene Revolution*). An alternative date for the start of the Pleistocene (and the *Quaternary Period*), which is in some ways more logical, is about 2.6 mya.

Lowestoft Glaciation: an alternative, now little used, name for the *Anglian Glaciation,* c.474,000–427,000 BP. It takes its name from the Lowestoft Till, laid down during this glaciation and found at the *type locality* on the Suffolk coast.

Lumbar vertebrae: the *vertebrae* of the lower back, below or (in quadrupeds) behind the rib cage.

Luminescence dating: see *Thermoluminescence* and *Optically Stimulated Luminescence* (OSL) dating.

Lyngby axe: an artefact (perhaps in fact used as a club) made from reindeer antler. The top of the antler and several of the tines are sawn off and then the antler beam is used as the handle and one of the tines as the 'blade'. Named after a find spot in Denmark. They are associated with the *Ahrensburgian* culture of Germany and the Low Countries, belonging to the *Terminal Upper Palaeolithic*. A large example has been found at Earls Barton in Northamptonshire.

Macrofossil: a *fossil* (plant or animal) large enough to be seen with the naked eye.

Magdalenian: a *Late Upper Palaeolithic tradition* widespread in Europe from around 22,000–14,000 *BP.* Associated in particular with abundant *mobiliary art*, including carved figurines and *batons*, and *parietal* art, including the famous cave paintings of Lascaux in France and Altamira in Spain. Named after La Madeleine rockshelter in the Dordogne. The *Creswellian* in Britain is a late variant of the Magdalenian, the links with the continent now being reinforced by the recent discovery of cave art at Creswell Crags.

Magnetic dating: see *geomagnetic reversals*.

Magnetic reversals: see *geomagnetic reversals*.

Mammal: a *taxon* of *vertebrate* animals (*Class* Mammalia) characterised by features such as hair, lactation and the presence of three middle-ear bones.

Mammal Assemblage Zone (MAZ): the formal name given to significant groupings of *mammals* typical of different periods of the British *Pleistocene*. These are based on the presence and absence of particular species, evolutionary developments amongst some lineages and other characteristics such as body size change. MAZs provide a very important *relative*, and through *cross-dating, absolute* dating method for Pleistocene deposits, and are particularly important for the dating of *river terrace* deposits. MAZs are named after *type localities*, such as the Swanscombe MAZ. Some MAZs correlate with particular *Marine Isotope Stages*, such as the Swanscombe MAZ (*MIS* 11) or the Purfleet MAZ (MIS 9), while others can define sub-stages, such as the Ponds Farm MAZ (early MIS 7) and the Sandy Lane MAZ (late MIS 7). See also *biostratigraphic dating* and *Vole Clock*.

Mammoth Steppe: a term used to describe the landscape of northern Eurasia and North America, from Britain to Alaska, during various phases of the *Pleistocene*. Without any parallel today, it comprised a mixture of steppe and tundra vegetation supporting an abundant *mammalian fauna*, most conspicuously the woolly mammoth. Also called *steppe-tundra* or *tundra-steppe*.

Mandible: lower jaw bone.

Marginal moraine: a *moraine* that forms at the margins of a *glacier*, including both *lateral* and *end moraines*.

Marine oxygen isotope record: the record of climate change inferred from cores drilled into the ocean bed (*deep sea cores*). The change in climate is inferred from changes through time in the relative proportions of oxygen *isotope* 16 (^{16}O) and the slightly heavier oxygen isotope 18 (^{18}O) in the sea water. The lighter isotope ^{16}O evaporates more easily than the heavier ^{18}O and in times when *ice sheets* expand globally more of the lighter isotope is 'locked up' in the glaciers, and it is relatively depleted in sea water. When the glaciers melt and the water locked up in them is returned to the oceans, the relative proportion of ^{16}O in sea water increases again. Measurement of the relative proportions of ^{16}O and ^{18}O in the oceans of the past is made by analysing their proportions in the shells of tiny marine organisms called *foraminifera*, which use the oxygen in the sea water to build their calcium carbonate shells, and which thus preserve the relative proportions of the heavy and light isotopes in the sea water at the time the organisms died. When the organisms die they sink to the sea floor and accumulate steadily in the thick deposits of sea floor sediment. Cores drilled into the sea-floor sediment recover a stratified sequence of the foraminifera, and analysis of the oxygen isotope proportions in their shells provides a *proxy* record of the growth and decay of the ice sheets, and thus of global climate, over many thousands of years. The changing proportions of the two isotopes are presented in the form of a graph – the oxygen isotope curve – in which warm peaks and cold troughs can be identified. One of the most recent curves combines data from 57 different cores, giving a continuous record of climate change covering the last 2 million years. Compare *ice cores*.

Marine Oxygen Isotope Stage (MIS): one of the numbered climate stages (even numbers for cold, odd numbers for warm) into which the *marine isotope record* is divided. The numbering system starts with MIS 1, the present-day warm *interglacial* stage, and works backwards. These stages are called oxygen isotope stages (OIS) in the older literature but as an oxygen isotope record is now also provided by *ice cores*, the term MIS is used to avoid confusion.

Marine regression: a fall in sea level causing exposure of land. The fall may be either absolute (see *eustasy*) or relative to the land in question (see *isostasy*).

Marine transgression: a rise in sea level causing flooding of land. The rise may be either absolute (see *eustasy*) or relative to the land in question (see *isostasy*).

Marl: an unconsolidated *sedimentary* rock intermediate between clay and *limestone*.

Matrix: in geology, the finer-grained material that surrounds larger grains in a rock or *sediment*.

Matuyama Chron: a period of (mainly) reversed geomagnetic polarity, when magnetic north and south were reversed (i.e. a magnetic compass would have pointed south) – see *geomagnetic reversals*. The Matuyama Chron lasted from about 2.59 *mya* to 0.78 mya, but contains several briefer periods when the polarity was normal. The transition from the Matuyama Chron to the *Brunhes Chron* of normal polarity is taken to mark the transition from the *Lower Pleistocene* to the *Middle Pleistocene*. Named after Motonori Matuyama (1883-1958), one of the discoverers of geomagnetic reversals.

Maxilla (plural **maxillae**): upper jaw bone.

Meandering river: single thread river characteristic of warm climates, like those seen in Britain today.

Megafauna: in the context of the *Pleistocene*, mammal species over 40 kg (90 lb) in weight (about the size of a wolf). Note however that the definition of megafauna is not consistent between authorities.

Member: in geology, a stratigraphic unit comprising several *beds* related spatially and in terms of their rock type.

Mesolithic: 'Middle Stone Age', coming between the *Palaeolithic* ('Old Stone Age') and the *Neolithic* ('New Stone Age). In Britain – and Europe more generally – the Mesolithic describes the hunter-gatherer societies of the early part of the current *interglacial* (*Holocene*) prior to the advent of farming societies, which mark the advent of the Neolithic.

Mesozoic: literally 'middle life', the geological *era*, about 251–65 million years ago, between the *Palaeozoic* ('ancient life') and the *Cenozoic* ('recent life'). The age of the dinosaurs.

Metacarpal bones: bones in the hand of a human, or the forelimb of other *mammals*, between the *carpal* (wrist) bones and the *phalanges* (fingers/toes).

Metamorphic rock: a rock that has undergone changes due to the effects of temperature and pressure after the rock first formed. The other two major rock types are *sedimentary* and *igneous*.

Metatarsal bones: bones in the foot of a human, or the hind limb of other animals, between the *tarsal* (ankle) bones and the *phalanges* (toes).

Micoquian: a late *Acheulean* stone-tool *industry* characterised by carefully shaped, elongated and pointed (lanceolate) *handaxes*. Named after the rock shelter of La Micoque in the Dordogne.

Microfauna: very small animals whose *fossils* must be identified under a microscope.

Microfossil: a *fossil* (plant or animal) too small to be seen with the naked eye, e.g. *pollen*.

Microlith: a small, standardised stone *flake* or *blade*, originally set into a composite tool or weapon (e.g. spear or arrow). Typical of the *Mesolithic* period.

Micromorphology: in biology, *geology* and soil science, the fine-level structure (*morphology*) of an organism, mineral or soil/sediment component, visible only with a microscope.

Micron: the length of 0.001 mm, generally written 1 µm. Used for describing the size of small particles, such as silt.

Microwear analysis: see *use-wear analysis*.

Middle Palaeolithic: the second of the three major divisions of the European *Palaeolithic*, dating to about 300,000–40,000 *BP*. It is recognised by the introduction of the *Levallois technique* (*Mode* 2) of stone-tool making and is associated in Europe with the Neanderthals.

Middle Pleistocene: the central sub-division of the *Pleistocene epoch*, about 780,000–127,000 *BP*. The transition from the *Lower Pleistocene* to the Middle Pleistocene is marked by the *Brunhes-Matuyama Reversal* at about 780,000 *BP*, while the transition from the Middle to *Upper Pleistocene* is marked by the beginning of the *Last (Brunhes) Interglacial* (*Ipswichian/Eemian*) at around 127,000 BP. See also *Middle Pleistocene Revolution*.

Middle Pleistocene Revolution (MPR): a name given to a change in both the frequency and the amplitude of *glacial/interglacial* cycles that occurred around the time of the transition from the *Lower* to the *Middle Pleistocene*. Between about 900,000 to 650,000 BP, the interval between each cycle became markedly longer (generally of the order of about 100,000 years), while the temperature difference between *glacial* and *interglacial* phases became substantially greater, including some very severe glacials. This change is associated with a shift in the orbit of the Earth around the Sun from a circular to an elliptical orbit. Alternatively called the Middle Pleistocene Transition (MPT).

Milankovitch cycles: cyclical changes in the Earth's orbit, and in the tilt and 'wobble' of the Earth on its axis, and the collective effect these are predicted to have on *insolation* and climate change on the Earth, notably the onset of glaciations in a similarly cyclical manner. Although the fit between *glacial/interglacial* cycles as recorded in the *deep sea cores* and the orbital periods is not perfect, the orbital theory has very wide support. Named after the Serbian mathematician Milutin Milankovitch (1879-1958) who did most to develop the orbital theory of glacial/interglacial cycles. See also *astronomical theory*.

Mindel Glaciation: the name given in the Alpine region to the *glacial* equated with the *Anglian* in Britain. Before the impact of the *marine isotope record*, the scheme of just four glaciations during the *Pleistocene* (*Günz, Mindel, Riss* and *Würm*) worked out in the north Alpine region in the early 20th century was used as a framework much more widely, including in Britain. It is now realised that this was a gross oversimplification.

Mindel-Riss Interglacial: name given in the Alpine region to the *interglacial* equated with the *Hoxnian* in Britain.

Mineralisation: the process by which *fossils*, while buried in the ground, gradually accumulate inorganic minerals that infill and/or replace the original bone, tooth or shell, making them very hard and dense.

MIS: see *Marine Isotope Stage*

Misfit valley: a large valley with a small stream, indicating that the valley was formed by a much larger Ice Age river.

Mobiliary art: portable art (e.g. figurines and pendants) produced during the *Upper Palaeolithic*. Compare *parietal art*.

Modes of lithic technology: system for classifying stone-tool technology proposed by the prehistorian Grahame Clark in 1969 and now widely used. While Mode 1 industries appeared earliest and Mode 5 industries latest, the aim is simply to describe the technological mode by which tools were made free from assumptions about date or cultural affiliation. For example, Mode 1 tools continued to be made long after the appearance of Mode 2 tools.

Mode 1: *Pebble tool* industries using *choppers* and simple *flakes* struck off pebbles.

Mode 2: *Bifacially* worked tools (*handaxes* and *cleavers*) produced from large flakes or *cores*.

Mode 3: Flake tools produced from *prepared cores*.

Mode 4: Punch-struck *blades* that may be *retouched* into various specialised tool types.

Mode 5: *Microlithic* components of composite artefacts, often *backed* or otherwise *retouched*.

Molars: the cheek teeth of *mammals*, adapted for grinding foodstuffs.

Molluscan analysis: the analysis of the remains of *molluscs* from deposits in order to reconstruct ancient climates and environments. Compare *insect analysis* and *pollen analysis*.

Molluscs: members of the large and diverse Mollusca *phylum*, including *gastropods* (snails and slugs) and *bivalves* (mussels and clams). Where conditions are not acidic, their shells are found extensively in *Pleistocene* deposits and they are used for palaeoenvironmental reconstruction.

Monolith tins: tins (generally about 50cm x 10cm x 10cm, with one long side open) which are hammered into a section (exposed face of a sedimentary sequence) in order to collect samples of *pollen*, seeds, *molluscs* and insects, and for *radiocarbon dating* and soil *micromorphology*. Several tins will be used together, hammered into the section in an overlapping manner, in order to obtain a continuous sample of the entire stratigraphic sequence. Also known as *Kubeina tins*.

Moraine: a recognisable landscape feature comprised of an accumulation of rocks and other debris deposited by a *glacier*.

Morphology: physical structure or form.

Mosaic vegetation: a vegetation comprising a wide range of plant species, often an amalgam of different ecological types.

Mousterian: a *Middle Palaeolithic* stone-tool industry associated with the *Neanderthals*. Mousterian assemblages include a variable number of tools made using the *Levallois* technique. Some of the common tool types include *Levallois points*, various kinds of scrapers, *burins, backed* knives, choppers, *denticulates* and *handaxes*. Named after the *type site* of Le Moustier, a rock shelter in the Dordogne.

Mousterian of the Achulean Tradition (MAT): a typological division of the *Mousterian* stone-tool industry. The French prehistorian François Bordes (1919-1981) divided the *Mousterian assemblages* of southwest France into five groups, each containing different proportions of tool types. He believed these to represent five different tribes but this is very unlikely, given the duration of the tradition over tens of thousands of years, and alternative explanations have been offered. The Mousterian tools found in Britain appear to belong to the Mousterian of the Achulean Tradition, one of Bordes' groups, and date to the British *Late Middle Palaeolithic,* about 60,000–40,000 BP. The most recognisable element in these assemblages is the *bout coupé handaxe*, and use of the *Levallois technique* is rare.

Munsell Soil Colour Chart: a chart, devised by Albert Munsell, used by soil scientists and others to enable accurate and objective description of the colour of soils, other sediments and rocks. The charts are contained in a small ring-bound book, and are based around the three critical factors of hue, value and chroma, all rated on a scale of 1–10.

Musk ox: a *species* of *bovid* (<u>Ovibos</u> <u>moschatus</u>; *Order Artiodactyla; Family* Bovidae) adapted to high arctic conditions. Characterised by a dense coat and large, curved 'helmeted' horns. Surprisingly the musk ox is a *caprid* (i.e of the sub-family capridae) and therefore more closely related to sheep and goats than to oxen. The male emits a strong odour to attract females during the mating season – hence the name. Now surviving only in northern Canada and Greenland, but with some recent reintroductions elsewhere, they inhabited Britain during some colder intervals of the *Ice Age*.

Mustelid: name for the 'weasel family' of *mammals* (from Latin 'mustela', weasel), belonging to the carnivore *order* (*Family* Mustelidae). Includes otters, badgers, stoats, polecats and the wolverine.

Mutual Climatic Range (MCR) method: a method for using the present-day climate tolerances of different *species* of beetles (and occasionally other animals) to infer past climates from *fossil assemblages*. First the present-day distributions of the species in a fossil assemblage are mapped and the climatic range of each species established using meteorological data. This allows the distribution of each species to be re-mapped in 'climate space' along two axes – temperature of the warmest month in which the beetle is found (T_{MAX}), and the temperature range between the warmest and coldest months (T_{RANGE}). Investigating the overlap between the 'climate space' occupied by a number of species then allows ancient temperatures to be calculated with a high degree of accuracy.

Mya: abbreviation for 'million years ago', e.g. 6 mya means 'six million years ago'.

Myr: abbreviation for 'million years', e.g. 6 myr means 'six million years'.

NAP: in *pollen diagrams,* an abbreviation for non-arboreal pollen, i.e. pollen from plants other than trees treated as a group. Compare *AP.*

Natural selection: the mechanism, famously first posited by Charles Darwin and Alfred Russel Wallace, that causes organisms to be adapted to their environment and lineages of organisms to change through time in response to changes in their environment ('descent with modification'). Natural variation between the individuals of a *species* means that some are, by chance, slightly better adapted to their environment than others and thus more likely to survive, reproduce and pass on their genes to their offspring ('survival of the fittest'). This results, over many generations, in evolutionary change and varied and sometimes breathtakingly complex adaptations ('endless forms most beautiful and most wonderful').

Neanderthal: see *Homo neanderthalensis*.

Negative bulb of percussion: the scar sometimes left on a *core* when a *flake* is struck from it, corresponding to the positive *bulb of percussion* on the struck flake.

Negative flake scar: see *flake scar*.

Neolithic: 'New Stone Age'. The archaeological period following the *Mesolithic* and associated with the introduction of agriculture and animal husbandry. In Britain this period lasted from about 6,500 *BP* to about 3,800 BP (4,500–1,800 BC). Followed by the Bronze Age.

Nitrogen test: a test which can be used to determine the date of a bone relative to others in the same deposit. Nitrogen is present in bone at about 4% but the level gradually declines after burial. This cannot be used as an *absolute* dating method because the rate of decline is affected by environmental factors such as temperature. However, comparing the level of nitrogen surviving in two bones from the same deposit can determine their relative date. The nitrogen test showed clearly that the Piltdown jawbone was recent, supporting results from the *fluorine test* in proving Piltdown Man as a hoax.

Notch: in stone-tool technology, a *denticulate* tool with a single indentation.

Numerical dating: see *absolute dating*.

Nunatak: the peak of a mountain projecting above the surface of an *ice sheet* or highland ice field (an Inuit word).

Occipital bun: a term used by anatomists to describe the pronounced bulge in the occipital bone (the bone at the back of the skull) seen in Neanderthal specimens.

Occlusal surface: the masticating (chewing) surface of *molar* and *premolar* teeth.

Ocean Conveyor Belt: see *thermohaline circulation*.

Older Dryas Stadial: a brief *stadial* (colder period), about 14,300–13,800 *BP*, during the *Lateglacial Interstadial*. In the record for continental northwest Europe, the Older Dryas intervenes between the *Bølling Interstadial* and the *Allerød Interstadial*. Both shorter and much less severe than the *Younger Dryas Stadial*. Named after the alpine/tundra flower *Dryas octopetala* (mountain avens), the pollen of which is characteristic of deposits of this episode.

Oldowan: name given to the oldest stone-tool industry, found in Africa from 2.6 *mya* and consisting of simple chipped *cores* and *flakes* (Mode 1). Named after the Olduvai Gorge in Tanzania.

Old Red Sandstone: name given to a suite of rocks laid down mainly in the *Devonian* period and found in Scotland, south Wales and the Welsh Marches. The red colour derives from the presence of iron oxide.

Optical dating: an alternative name for *Optically Stimulated Luminescence (OSL) dating*.

Optically Stimulated Luminescence (OSL) dating: a dating technique applied to minerals such as quartz and feldspar found in Pleistocene *aeolian* and *fluvial* deposits, which typically occur in sand and gravel quarry contexts. The technique measures how much natural background radiation buried grains of these minerals have accumulated since they were last exposed to sunlight, which clears radiation accumulated earlier. This provides an estimate of how long the grains have been buried, and thus the date at which the deposit was laid down and the date of deposition of any fossils or artefacts associated with the deposit. In the laboratory the grains in a sample are made to emit their stored radiation in the form of light (luminescence), and this is measured to estimate the amount of time that has elapsed since burial. The technique is most typically used to date sediments up to 400,000 years old, though refinements to method continue to push back this limit.

Orbital forcing: the effect of changes in the orbit of the Earth, and of changes in the Earth's tilt on its axis, on climate. See also *Milankovitch cycles* and *astronomical theory*.

Orbital tuning: adjustment of the chronology of events recorded in the geological or climatic record (e.g. the temperature variations extrapolated from *deep sea cores*) to fit in with the expected pattern predicted by *Milankovitch cycles*. In other words, 'tuning' variations in the deep sea or terrestrial record so that they are in sync with *astronomical theory*.

Order: the major *taxonomic rank* in the *Linnaean system* or hierarchy that is below *class* and above *family*. For example humans belong to *hominid* family in the *primate* order, which is one of several orders in the mammalian class.

Ordovician: the geological *period* dating to about 488–444 million years ago.

Organic: relating to biological material or processes, such as the remains of plants and animals.

OSL dating: see *Optically Stimulated Luminescence dating*.

Osteocalcin sequencing: a technique for measuring the degree of genetic relatedness, and hence evolutionary relationships, between animal *fossil* specimens. Ostecalcin is the second most abundant protein in bone after collagen. As a result of genetic mutation the structure of the osteocalcin molecule changes through time. By measuring the degree of similarity between the osteocalcin sequences in different specimens the degree of genetic relatedness between the specimens can be estimated.

Ostracods: small *bivalve* crustaceans, some of which live in salt water and others in fresh water. Used in environmental reconstruction as *proxys* for temperature, salinity and sea- and lake-level change.

Outwash: see *glacial outwash*.

Oxbow lake: a curved lake found on the *floodplain* of a river ('bayou' in America). They form when a loop of the river's meander is cut off and the river adopts a shorter, straighter course.

Oxygen isotope analysis: analysis of the relative proportion of the light oxygen *isotope* ^{16}O to the heavier oxygen isotope ^{18}O in the ice layers of ice sheets or in the shells of *foraminifera* in sediments on the sea bed. The changing proportion of these isotopes through time provides a *proxy* measure of glacier size and sea levels in the past, and thus of past climates. See *marine isotope record* for more detail. See also *ice cores*.

Pachyderm: an informal term for a thick-skinned *quadruped*, such as the elephant, rhinoceros or hippopotamus.

Pakefield Interglacial: an *interglacial* recognised in deposits of the *Cromer Forest-bed Formation* and tentatively identified with the *MIS* 17 interglacial (c. 712,000–659,000 *BP*). At the time of writing, artefacts (*flint flakes* and *cores*) from deposits at Pakefield associated with this interglacial – dated to around 700,000 BP by a variety of methods – provide the earliest confirmed record of human occupation in Europe north of the Alps. A rich range of *proxies*, including insects, *molluscs*, plants, *mammals* (including hippo) and *soil carbonates*, suggest a Mediterranean-type climate in Britain at the stage of the interglacial when the artefacts were deposited. Named after the village of Pakefield on the Suffolk coast.

Palaeo-: 'old' or 'ancient', used in combination with other words.

Palaeoanthropology: the study of fossil humans and related species.

Palaeochannel: an ancient infilled stream channel no longer carrying water and often deeply buried under alluvium. Palaeochannels may be a rich source of *fossil* and *artefact* remains.

Palaeoenvironmental: relating to the ancient environment.

Palaeolith: literally 'old stone'. A name given to any stone *artefact* of the *Palaeolithic* period.

Palaeolithic: the 'Old Stone Age', covering the period from the earliest known stone tools (from Africa c. 2.6 million years ago) to the beginning of our present *interglacial* c. 11,400 years ago. In Britain the earliest stone tools are presently dated from around 700,000 years ago.

Palaeomagnetic dating: a range of dating techniques that make use of known changes in the Earth's magnetic field to provide *absolute dates* for events. The technique most relevant to the *Pleistocene* is that which makes use of *geomagnetic reversals*.

Palaeontology: the study of *fossils*.

Palaeopathology: the study of disease in antiquity, both human (usually) and animal.

Palaeosol: a buried old soil.

Palaeozoic: literally 'ancient-life', the geological *era* between about 542–251 million years ago.

Palynology: the study of pollen and spores.

Parietal art: a term that literally means 'art on the walls'. In a *Palaeolithic* context it usually means art engraved or painted on the walls or ceilings of caves, although it can refer to art on any immovable surface.

Particle size analysis: systematic, quantitative analysis of the sizes of grains in a sediment. Particle size, from pebbles down to clay, in a sediment reflects how the sediment was formed and also subsequent alteration, e.g. through weathering.

Patina: a thin coating that may develop on the surface of stone tools, particularly *flint*, during burial. Common colours include yellow, cream and blue. Recent damage to patinated stone tools can be detected as the damaged area will not be of the same patination as the rest of the artefact.

Pebble tools: simple stone tools made by striking one or more *flakes* from a pebble.

Pedogenesis: the process of soil formation.

Pedology: soil science.

Penknife point: in stone-tool typology, the name given to a distinctive type of curve-*backed* point, with additional basal *retouch*. Dated to the *Final Upper Palaeolithic*, roughly 14,000–12,900 BP. British penknife points are closely related to German *Federmesser* points.

Percussion cone: in stone-tool technology, an alternative term for *cone of percussion*.

Percussion features: the characteristic features found on stone tools that reveal human manufacture, for example the *bulb of percussion, striking platform, negative flake scars* and *retouch*.

Percussion flaking: the removal of *flakes* from a *core* or tool through a hitting or striking action. Percussion flaking may be done with a *hard hammer* (stone) or a *soft hammer* (bone or antler). Compare *pressure flaking*.

Percussor: an alternative name for the *hammer* (a hard stone or softer bone or antler object) used in *knapping*.

Periglacial: describes land, climate, environmental conditions, formation processes, etc. close to an *ice sheet*, and therefore dominated by cold. *Permafrost* and *tundra* vegetation are typical of periglacial regions.

Périgordian: an *Upper Palaeolithic* stone-tool *tradition* in southwest France. Only the Upper Perigordian is now generally recognised, which is essentially equivalent to the *Gravettian*.

Period: in geological terminology, 'period' has the specific meaning of an interval of time that is a subdivision of an *era*. For example the *Quaternary* period is a subdivision of the *Cenozoic era*. *Epochs* are subdivisions of periods. For example the *Pleistocene* epoch is a sub-division of the Quaternary period. However, 'period' is very widely used, in this book as elsewhere, in the looser sense of meaning any interval of time.

Perissodactyla: the *order* of *mammals* comprising the odd-toed *ungulates*. Includes horses (one toe) and rhinos (three toes). Compare *Artiodactyla*.

Permafrost: permanently frozen ground, which may be hundreds of metres thick. Only the top few metres (the *active zone)* may temporarily thaw in summer.

Petrography: in geology, the sub-discipline concerned with the description, identification and classification of rocks.

Petrological microscopy: the study of *thin sections* of rocks for the purpose of identification and location of geographical source.

Petrology: similar to *petrography* but broader in scope and including the origin and history of rocks.

Phalange: finger or toe bone. The phalanges of the hand in humans or the fore limb in other mammals are called manus phalanges, while those of the leg or hind limb are called pes phalanges.

Photomicrograph: a photograph taken at high magnification in order to reveal details of form and structure, for example of a pollen grain.

Phylogeny: the evolutionary relationships between *species* in terms of their ancestor/descendant relationships, often conceived of as the 'tree of life'.

Phylum (plural **phyla**): in biology, the major *taxonomic* rank in the *Linnaean system* of classification that is below *kingdom* and above *class*. For example, the *mammals* are one of several classes within the chordate phylum, which in turn is one of several phyla within the animal kingdom.

Physical anthropology: the branch of *anthropology* that studies the evolution of humans, their variability and adaptability. Also called biological anthropology. Similar, but somewhat broader in scope, to *palaeoanthropology*, which is primarily concerned with the fossil record.

Phytoliths: minute particles of silica derived from the cells of plants and able to survive long after the plant has decomposed. Traces of phytoliths on stone tools provide a clue to their former use. See also *residue analysis*.

Plankton: drifting organisms that inhabit the near-surface zone of the oceans. They may be plants, animals or bacteria, and are defined by their ecological niche rather than any genetic similarities. Compare *benthic organisms*.

Plano-convex: in stone-tool technology (especially), flat on one side (usually the *ventral* surface) and curved on the other (*dorsal* surface)

Plate tectonics: the movement of large plates of the *lithosphere* acting as rigid slabs floating on a viscous mantle, and causing such processes and structures as ocean-floor spreading, *continental drift*, mountain building, earthquake zones and volcanic belts.

Platform preparation: in stone-tool technology, any adjustment that is made to the *striking platform* to prepare it for *flake* removal.

Pleistocene: the geological *epoch,* formally dated from 1.8 *mya* to 11.4 kya, that is the earlier of the two subdivisions of the *Quaternary period*, the later sub-division being the *Holocene* (our present *interglacial*). The formal name for the *Ice Age*, as used in this book. An alternative date of 2.6 mya for the commencement of the Pleistocene/Quaternary/Ice Age is in many ways preferable but is the subject of controversy.

Pliocene: the geological *epoch,* dated from 5.3 to 1.8 *mya,* that precedes the *Pleistocene*. The early evolution of the *hominins* (genera such as Australopithecus) took place in Africa during the Pliocene, with the evolution of the genus *Homo* and its expansion out of Africa taking place at the Pliocene-Pleistocene boundary.

Point bar: a low ridge of *sediment* that forms along the inner bank of a meandering stream.

Point of percussion: in stone tool manufacture, the place on the *butt* of a *flake* where the *hammer* blow was struck which removed the flake. When *hard hammer* percussion is used, a distinct *cone of percussion* will form underneath the point of percussion.

Pollen: pollen consists of grains which are the male gametes of flowering plants. Pollen grains survive very well in some conditions, notably *anaerobic* and acidic deposits and soils, and are used to reconstruct Ice Age environments.

Pollen analysis: the study of *pollen* that has been preserved in ancient deposits in order to reconstruct past environments. The pollen is extracted, identified under a microscope, and counted. The relative abundance of the various groups present is then used to reconstruct former vegetation. Pollen analysis will often involve the study of pollen from a stratified sequence of deposits in order to provide a reconstruction of changes in vegetation through time. Such stratified sequences may be extracted from an exposed section, for example using *monolith tins,* or may be obtained using various coring techniques. The changes in relative abundance of different pollen groups through time is displayed on a *pollen diagram*.

Pollen diagram: a diagram used to display the changing proportions of different *pollen* types through time, usually as derived from a single stratified sequence of deposits. As well as showing the changing proportions of different pollen *genera* (e.g. birch, elm, pine, oak, etc.) through time, the diagram will often show the relative proportions of arboreal pollen (AP) and non-arboreal pollen (NAP), and may divide the sequence up into *pollen zones*.

Pollen zones: the changing assemblages of *pollen* through the course of an *interglacial* have been divided up into pollen zones. Each zone contains a pollen assemblage typical of a particular stage in the development of vegetation during the course of an interglacial. Four principal zones have been recognised: Zone I 'Pre-temperate zone' (rising values for arboreal pollen, particularly of *boreal* trees such as birch and pine. Light-demanding shrubs and herbs are initially dominant, but rapidly decline); Zone II 'Early temperate zone' (dominated by trees associated with mixed-oak woodland, such as oak, elm and hazel, with non-arboreal pollen minimal); Zone III 'Late temperate zone' (expansion of some trees not present in Zone II, e.g. alder and spruce, and gradual disappearance of others); Zone IV 'Post-temperate zone' (boreal trees dominant once again, rising values of non-arboreal pollen). It is now recognised that this scheme greatly oversimplifies matters, and the scheme does not always fit well with either the climate evidence from the *oxygen isotope record* or the evidence from *mammal assemblage zones*.

Positive feedback: any process that tends to amplify change once change has been initiated. A classic example of positive feedback is the *albedo* effect of snow cover. Snow cover reflects solar radiation, leading to cooling. The cooling leads to an expansion of snow cover, which leads to more reflection of solar radiation, which leads to more cooling, which leads to further expansion of snow cover, and so on.

Postcranium: the skeleton below the head.

Postglacial: an alternative term for the *Holocene*, the present-day *interglacial epoch*.

Post-glacial rebound: an alternative term for *isostatic uplift* or *glacial rebound*.

Potassium/Argon (K-Ar) dating: a *radiometric* dating technique used to date volcanic rocks no more recent than about 100,000 years old. It is based on the very slow decay of the radioactive *isotope* potassium-40 (^{40}K) to the inert gas argon-40 (^{40}Ar) in the rock. The technique is not applicable to *Pleistocene* Britain due to the lack of appropriate deposits but has played a major role in the dating of *hominin* sites in the volcanic landscapes of the Rift Valley in Africa (e.g. Olduvai Gorge). A more sensitive variant of the method is 'argon-argon' dating.

Precession (also called **precession of the equinoxes**): the phenomenon whereby the seasons, over time, occur at different times of the calendar year. This is caused by the Earth's slow wobble on its axis (like a child's spinning top), each full revolution of the wobble taking about 27,000 years to complete. Precession is thought to be an important *orbital forcing* mechanism of cyclical climate change, through causing redistribution of *insolation* over the northern hemisphere.

Premolars: 'transitional' teeth located between the *canines* and *molars*.

Prepared Core Technology (PCT): in stone-tool making, this describes techniques in which a *core* is carefully prepared so that a *flake* or *blade* (or a series of flakes or blades) of predetermined shape can be struck from it. The widespread adoption of PCT marks the transition from the *Lower Palaeolithic* to the *Middle Palaeolithic*. The *Levallois technique* is an example of PCT.

Pressure Flaking: the removal of small *flakes* from the edge of a tool by pressing (rather than hitting as in *percussion flaking*) with a small piece of antler or bone.

Primary context: *artefacts* or *fossil* remains are said to be in a primary context when they are found more or less where they were discarded or where the animal died. Primary context sites are very rare because numerous factors, for example the action of rivers, normally serve to move artefacts and fossils from their original resting place and redeposit them in a *secondary context* somewhere else.

Primates: the biological *order* within the *mammalian class* that comprises the prosimians (lemurs, bush babies, etc.), the monkeys and the apes, the latter including humans. Distinctive characteristics include forward-facing eyes, a large brain and a long childhood.

Proboscidea: the *order* of *mammals* that includes the mammoth and living elephants. Characterised (except for the earliest forms) by possession of a trunk (proboscis) and *tusks*.

Proglacial features: features that are related to *glaciers* but which are formed beyond the glacier itself, for example proglacial lakes.

Protozoa: single-celled organisms such as *foraminifera*, radiolarians and dinoflagellates.

Proximal: near. The proximal end of a long bone is the end nearest the body. In stone-tool technology, the proximal end of a *flake* or *blade* is the end that bears the *striking platform* and/or the *bulb of percussion*.

Proxy: in *climate* and environmental reconstruction, a proxy is something that provides indirect evidence for an important aspect of the climate or environment. For example, beetle fossils are a proxy for past temperatures, while *oxygen isotope* ratios are a proxy for the past extent of *ice sheets*.

Punch technique: in stone-tool making, this describes an *indirect percussion* method associated with systematic *blade* production. An antler, wood or bone punch is placed at the edge of the *striking platform*, the punch is then struck with a *hammer* and a long thin blade is detached from the core.

Purfleet Interglacial: name sometimes given in Britain to the *interglacial* corresponding to *MIS* 9 (about 334,000–301,000 *BP*). Also known as Purfleet/Grays Interglacial. Named after the *type locality* on the Lower Thames.

Quadrupedal: walking on four legs. Compare *bipedal*.

Quartzite: a hard *metamorphic rock* transformed through heat and pressure from sandstone. During the Ice Age quartzite cobbles were used as *hammerstones* for *flint knapping*, and in regions without *flint* were also used as the raw material for stone tools.

Quaternary: literally the 'fourth', or most recent, *period* of geological time, dated from either 2.6 *mya* or 1.8 mya (a matter of academic controversy) to the present. Sub-divided into the *Pleistocene* (Ice Age) and *Holocene* (present-day *interglacial*) epochs.

Quaternary science: the multi-disciplinary study of the *Quaternary period*. Broader disciplines that contribute to Quaternary science include *geology, geomorphology*, geography, *archaeology*, soil science, palaeobotany, *palaeontology* and palaeoclimatology.

Radial fissures: in stone-tool technology, small cracks or lines that radiate out from the *bulb of percussion* on a *flake*. They are also sometimes referred to as 'hackles'.

Radioactive isotopes: unstable *isotopes* of atoms that change into other atoms by emitting charged particles. This process occurs at known rates for different isotopes and can therefore be used for geological and archaeological dating purposes.

Radiocarbon dating: a *radiometric* dating method based on the decay of carbon-14 in dead organisms, pioneered by Willard Libby in the late 1940s. It can be applied to most organic materials and can provide dates up to approximately 45,000 years ago, although *calibration* beyond about 11,000 *BP*, the end of the Palaeolithic, presents difficulties. See also *accelerator mass spectrometry* and *ultrafiltration*.

Radiometric dating: a means of dating rocks, minerals, sediments and fossils by means of *radioactive isotopes*.

Radius: in anatomy, one of the two bones of the lower arm (with *ulna*).

Raft: in *Quaternary geology,* a raft (or glaciotectonic raft) is a slab of rock or unconsolidated sediment that has been removed from its source and transported by a *glacier*.

Raised beach: beach deposits found inland, above present-day sea level, representing an episode of *marine transgression*.

Reduction: in stone-tool technology, this term describes the process of creating an artefact by means of flaking material from a blank. The whole process is called the reduction sequence. An example is the reduction of a flint nodule to create a *handaxe*.

Refitting: in archaeology, the process of putting back together stone *flakes* struck from a *core* in order to recreate the original core, thus giving insights into the *knapper's* craft.

Refugium (plural **refugia**)**:** an area where some plants and animals (including *hominins*) survived during unfavourable climatic conditions, generally a glaciation, while others went locally extinct. Following climatic amelioration, the plants and animals generally expanded from the refugium to recolonise areas that had formerly been abandoned.

Rejuvenation: in stone-tool technology, refers to adjustment of a used *blade core* in order to prepare it again for controlled blade removal.

Relative dating: techniques of dating which enable deposits, *fossils* or *artefacts* to be put into a sequence relative to one another, but which do not in themselves provide *absolute dates* (calendar dates). The principal relative dating technique is stratigraphy, whereby deposits are arranged in a sequence based on their physical relationships, with later deposits typically overlying earlier ones. The evolutionary progression of fossils and the typological development of artefacts are also used to provide sequences of relative dates. Often such sequences can be tied in to absolute chronologies by applying appropriate absolute dating methods to material in the sequence, or by *cross-dating* a sequence which has not been dated by absolute methods to one that has.

Residue analysis: in the study of stone-tools, residue analysis involves the identification of plant (e.g. *phytoliths*) or animal (e.g. blood) residues that may survive on stone artefacts. Identification will normally involve using a microscope and/or chemical analysis. The residues which sometimes survive on stone tools provide an important clue to their use, along with *use-wear analysis*.

Retouch: in stone-tool making, the secondary modification of the edge of a *flake* or *blade* through the removal of small flakes by means of *percussion* or *pressure flaking*. The removal of small flakes makes these edges more durable. Retouch removals can be taken from one face of an edge (unifacial retouch) or both faces (bifacial retouch), and can be steep or shallow. Common retouched tools include *scrapers*, *notches* and *denticulates*. Retouch can also be used to *back* or blunt the edge of a blade for *hafting* or ease of handling.

Ripples: in stone-tool technology, shock ripples on a *flake* reflecting the transfer of the *hammer* blow energy through the stone as the flake is detached. They are best seen on fine-grained materials such as *flint*. The ripples radiate out latitudinally from the *striking platform* and *bulb of percussion*. They can also be seen in the *negative flake scars* on *cores*.

Riss Glaciation: the name given in the north Alpine region to the glaciation that is the temporal equivalent of the *Wolstonian* stage in Britain. However, the *marine isotope record* suggests that this 'stage' in fact includes three separate glacial episodes (*MIS* 10, MIS 8 and MIS 6).

Riss-Würm Interglacial: the name given in the north Alpine region to the interglacial between the *Riss* and *Würm Glaciations* corresponding to *MIS* 5e (the *Ipswichian Interglacial* in Britain), about 127,000–115,000 *BP*.

River terrace: part of an old *floodplain* that has been left perched on the side of a valley when a river has cut down through the old floodplain to a lower level. When this process has been repeated several times this will result in the creation of 'terrace staircase' down the side of the valley, the highest terrace being the oldest. Each terrace is typically the creation of a single *glacial-interglacial* cycle.

Rock flour: fine particles of rock created by glacial erosion and then carried away from the *glacier* in meltwater streams. Rock flour (also called glacial flour) may be carried long distances in suspension in rivers or by the wind, in the latter case being deposited as *loess*.

Rolled: in stone-tool typology, an artefact is described as rolled if its edges have been damaged by transport in a river.

Rose diagram: a diagram used to show the orientation of the long axes of *clasts*, etc.

Sagaie: a bone, antler or ivory point, thought to be the tip of spear.

Scapula (plural **scapulae**): shoulder blade.

Scraper: a stone tool with a convex working edge modified by *retouch*. Used for the preparation of hides and the shaping of wood, bone, antler etc.

Secondary context: *artefacts* and *fossils* are said to be in a secondary context when their findspot is different from their original place of deposition, usually due to movement by natural processes, especially transport by rivers. The majority of *Lower Palaeolithic* artefacts in Britain are found in secondary contexts in river valleys.

Sediment: material settled from suspension or solution in water, or material that has been transported by water, wind, ice or gravity, and has come to rest.

Sedimentary rocks: rocks formed by the compression of sediments, comprising both mineral and *organic* material, laid down by water, wind, ice, air or gravity. Examples of sedimentary rocks include sandstone, *limestone* and chalk. Compare *metamorphic* and *igneous* rocks.

Sedimentology: the scientific study of *sediments* and *sedimentary rocks*.

Shaft-wrench: see *baton*.

Shock ripples: see *ripples*.

Sidereal dating: an alternative term for *absolute dating*.

Sill: in geology, a *bed*-shaped *igneous* intrusion which is concordant with (i.e. runs between) the host beds. Compare *dyke*.

Silurian: the geological *period* from about 444 to 416 million years ago.

Slindon Raised Beach: *raised beach* deposits (the Goodwood-Slindon *Formation*), approximately 40m above present sea level, dating to an interglacial late within the *Cromerian Complex* (*MIS* 13, c. 528,000–474,000 *BP*) and defining the northern extent of the present-day Sussex Coastal Plain between Arundel and Westbourne, a length of 26 km. The famous *Lower Palaeolithic* site of Boxgrove, today about 10 km inland, is situated on the Slindon deposits.

Slopewash: a type of *sediment* formed by soil and rubble being moved down hill by such mechanisms as gravity, *solifluction* and washing by rain (but not transport by streams). The resulting deposit is usually characterised by poor *sorting* of the particles. An alternative term for *colluvium*.

Soft hammer: a type of *hammer* used in stone-tool making. Soft hammers can be made of bone, antler, wood or soft stones and tend to produce thinner flakes than hard hammers. This is useful for the final stages of shaping tools such as *handaxes*.

Soil carbonates: carbonate (*limestone*) nodules, about 2–3 cm long, that form in soils only under conditions of low rainfall or very seasonal rainfall. When found in fossil soils they therefore provide a *proxy* for past precipitation. *Stable isotope analysis* of soil carbonates provides further information about the conditions under which they formed – the ratio between oxygen *isotopes* ($^{16}O:^{18}O$) is controlled mainly by temperature, whilst the ratio of carbon isotopes ($^{12}C:^{13}C$) is controlled by the composition of the overlying vegetation.

Solifluction: the process of mass movement of water-laden soil and *sediment* as a result of the thawing of frozen ground.

Solutrean: an *Upper Palaeolithic culture* or *tradition,* mostly restricted to continental Europe west of the Rhône, characterised particularly by finely worked *bifacial* points produced by *pressure flaking*, but also associated with many artefacts made of bone and antler and an abundance of representational objects. Dates from around 26,000–22,000 BP, the earlier part of the *Last Glacial Maximum*, a time when Britain was abandoned. Where it is found the Solutrean seems to represent a significant break with the preceding *Gravettian* but elsewhere the Gravettian continues as the *Epigravettian*. Named after the *type site* of Solutré, near Macon, Saône-et-Loire.

Sorting: in the description of *sediments,* sorting refers to the extent to which the particles are arranged into layers of predominantly one size or not. This provides an indication of the processes that created the sediment. For example, rivers flow at variable strengths and will often sort sediments into distinct layers of different particle size, such as clays, silts, sands and gravels. Such deposits are described as 'well sorted'. Material deposited by a *glacier*, on the other hand, will be jumbled up or 'poorly sorted'.

Species: the most fundamental *taxon* in the *Linnaean system*, corresponding to what is generally perceived of as a 'natural kind'. The most common biological definition is that a species comprises a population of individual organisms sufficiently similar to each other to be able to breed and produce fertile offspring. Species are grouped into *genera* and are named using the 'binominal nomenclature' – see *Linnaean system*.

Speleothems: secondary mineral deposits formed in caves, typically *limestone* caves, by the precipitation of minerals from water seeping through cracks. Stalactites and stalagmites are examples of speleothems. Such deposits, which may seal *fossil* or archaeological remains, may be dated by *uranium-series* dating.

Spore: the asexual reproductive cell of non-flowering plants such as mosses and ferns.

Stable isotope analysis: the scientific analysis of the relative proportions of different stable (as opposed to radioactive) *isotopes* of various atomic elements in a range of substances. The relative proportions of stable isotopes can be used as a *proxy* for things as diverse as ancient climate and ancient diet. The relative proportions of the two stable isotopes of oxygen, ^{16}O and ^{18}O, in microscopic sea shells and in the layers of ice of a glacier can be used to infer the past climate and the extent of glaciers – see *marine isotope record* and *ice core record*. The ratios of stable isotopes of carbon ($^{12}C:^{13}C$) and nitrogen ($^{14}N:^{15}N$) in tooth enamel and bone collagen, of both humans and other animals, can be used to infer diet. For example, measuring the amounts of ^{13}C and ^{15}N in Neanderthal bones from Europe has indicated an almost exclusively carnivorous diet for the samples tested.

Stadial: a period of cold climate during an *interglacial* which is, however, neither long enough nor pronounced enough to be considered a *glacial*. Compare *interstadial*.

Stage: in geochronology, a stage is a subdivision of an *epoch*. Thus the *Pleistocene* epoch is divided into a number of named stages, usually named after regional *type localities*. Thus *Devensian* is the name of the last stage of the Pleistocene epoch in Britain. However 'stage' is also the term used for the various divisions of the marine isotope curve – *marine isotope stages* – and the Devensian encompasses several marine isotope stages (5d–2). Furthermore, the term is used informally, in this book as elsewhere, to denote any period of time.

Steppe: a flat landscape dominated by broad expanses of grassland and other herbaceous plants but devoid of trees. During much of the *Pleistocene* this kind of landscape dominated mid-latitude Europe.

Steppe-tundra: an alternative term for *Mammoth Steppe*.

Strangulated blade: in stone-tool technology, a *blade* in which the mid-region has been thinned to create two concave edges, forming a 'waist'.

Striking platform: in stone-tool technology, the place where the *hammer* strikes the *core* to detach a *flake*. Striking platforms can be natural or formed of old *flake scars* and may be prepared to alter the angle at which the hammer strikes the core, and thus modify the shape of the flake to be detached.

Subfossil: a term sometimes used to describe the ancient remains of animals or plants which have not become fully fossilised (i.e. mineralised). By this definition most *Pleistocene* remains are subfossils rather than *fossils*, but the term is seldom used.

Suids: members of the pig *family* (Suidae).

Swanscombe Interglacial: an alternative, older name for the *Hoxnian Interglacial*, both being names used in Britain for the *MIS 11 interglacial*, about 427,000–364,000 BP.

Taiga: the coniferous forest zone of northern latitudes, with a harsh continental climate, between the *tundra* and the *steppe*.

Tallies: enigmatic bone *artefacts* of *Upper Palaeolithic* date, often ribs, with notches or scratches carefully and regularly incised along their edges, sometimes in groups (some tallies from France have more elaborate markings). The notches are interpreted as tally marks, and it is argued that some tallies may have been used as calendars.

Talus: the scree, usually comprised of angular *clasts*, at the foot of a cliff or steep slope.

Tang: the tapered end of a projectile point, knife, etc. that fits into the shaft or handle.

Taphonomy: study of the processes by which animals, plants and *artefacts* are incorporated into *fossil* or archaeological deposits, especially the factors affecting the composition and completeness of excavated remains.

Tarsal bones: the bones of the human ankle or the equivalent bones in the hind limb of other *mammals*.

Taurodontism: a condition found in *molar* teeth in which the body of the tooth and the pulp chamber is enlarged, resulting in shortened roots. The condition is frequent in *Neanderthal* teeth.

Taxon (plural **taxa**): a unit of biological classification, such as *species, genus* or *family*.

Taxonomy: the branch of a science – chiefly biological science – concerned with classification.

Tectonics: in geology, the branch of study concerned with the forces and movements that have built geological structures, for example mountains.

Tephra: volcanic material, especially ash, transported by air (not lava).

Tephro-chronology: a chronology based on correlating volcanic ash beds.

Terminal moraine: a *moraine* at the end (snout) of a *glacier* that marks the glacier's furthest extent.

Terminal Upper Palaeolithic: the very last phase of the *Upper Palaeolithic,* after the cold peak of the *Younger Dryas Stadial,* roughly about 12,500–11,400 BP. Characterised by in Britain by long blade assemblages, including 'bruised blades' with distinct battering damage along their edges, perhaps resulting from use for chopping material such as antler.

Terrace: see *river terrace*

Tertiary: in older geological terminology the 'third period' of Earth history, preceding the *Quaternary*. In modern terminology the Tertiary equates with the first two *periods* of the *Cenozoic era,* the Palaeogene and the Neogene, spanning the time range from about 65 million years ago to about 2.6 million years ago.

Thermohaline circulation (THC): the system of large-scale ocean currents that play a key role determining the Earth's climate pattern. The adjective 'thermohaline' derives from 'thermo', meaning heat, and 'haline', referring to salt content. These are the two key variables that determine the density of sea water – colder, saltier water is more dense than warm, fresher water and will sink below it. The thermohaline circulation is also called the *Ocean Conveyor Belt*, which captures the idea of continuously circulating belts of warm surface- and cold deep-water currents. In the Atlantic, for example, the wind-driven Gulf Stream moves warm surface water from the Gulf of Mexico north-eastwards across ocean, cooling on the way and becoming more dense, until it eventually sinks at high latitudes (around 60° N), releasing enormous quantities of heat into the atmosphere, to form the 'North Atlantic Deep Water'. The return belt of the conveyer transfers this deep, dense cold water to the Southern Oceans. The effect of the Gulf Stream is to warm the climate of Britain by about 3°C above what would otherwise be expected. It is believed that this conveyor belt can be 'switched off', with profound effects on climate, for example by the release of huge quantities or fresh water into the Atlantic resulting from the collapse of the *Laurentide Ice Sheet*. This possibly happened during the *Younger Dryas Stadial*.

Thermoluminescence (TL): The emission of light by irradiated crystals when heated. Naturally occurring radiation dislodges electrons within crystal structures, the number of dislodged electrons accumulating over time. However, heating of the crystals releases the trapped electrons, the energy being emitted as light. This phenomenon is the basis of several absolute dating methods. Initial heating may be caused by fire, as the case of burnt flints or the firing of pottery, or sometimes simply by exposure to sunlight, as in the case of quartz grains in sand. This results in emptying of the electron traps in the crystal structure. After burial, natural radiation in the surrounding soil causes the gradual accumulation of electrons in the traps once again. The light released on reheating the sample in a laboratory provides a measure of the time that has passed since the initial heating event. See also *Optically Stimulated Luminescence (OSL)*.

Thinning flake: a by-product of *handaxe* manufacture. Characteristic flattish flakes removed from a handaxe using a *soft hammer* in order to finish or thin down the handaxe to the required shape for use.

Thin sectioning: the removal of a thin slice from a stone for the purpose of examination under a *petrological* microscope.

Thoracic vertebrae: the *vertebrae* of the upper back, along the length of the rib cage.

Tibia (plural **tibiae**): shine bone – the larger of the two bones of the lower leg. Compare *fibula*.

Till: the debris deposited directly by a *glacier*. This will normally be an un*sorted*, unconsolidated, non-stratified *sediment* (a *diamicton)* consisting of a wide range of particle sizes from larger *clasts* such as boulders, cobbles and pebbles, down to finer material such as sand, silt and clay. This mixture is why till is sometimes referred to as *boulder clay*.

Toolstone: raw material used to make stone tools.

Tortoise core: in the *Levallois* technique of stone-tool making, the prepared *core* from which a *flake* or flakes of predetermined shape is struck is often of a characteristic shape reminiscent of the shell of a tortoise.

Trace fossil: features found on surfaces or within *sediments* that were made by the activity of organisms, such as animal tracks and burrows, rather than the remains of the organisms themselves.

Tradition: with reference to stone tools, a tradition is an *industry* or a group of related industries found throughout an extended period of time, amounting at least to several generations. In practice the distinction between an industry and a tradition is not clear cut other than the emphasis on time depth in the latter, and the terms are often used more or less interchangeably (including in this book). Traditions are often named after a type site. The *Acheulean*, *Clactonian* and *Mousterian* are all examples of stone-tool traditions. Compare *culture*.

Tranchet: a chisel-ended stone tool, often a *handaxe*, with a straight, sharp working edge produced by striking off a flake at right angles to the main axis of the tool.

Transgression: see *marine transgression*.

Triassic: the geological *period* from about 251 to 200 million years ago.

Tribe: in the *Linnaean system* of biological classification, 'tribe' is a rank in the hierarchy that comes between *genus* and *family*. The genus Homo belongs to the *hominin* tribe (hominini) within the *hominid* family (hominidae). The other tribe within the hominid family is the panini (chimpanzees).

Trichoptera: caddis flies. Small moth-like insects with an aquatic larval stage. They are sensitive indicators of water conditions and may be used in palaeoenvironmental reconstruction.

Tufa: a *calcite* rock that forms when calcium carbonate is precipitated from lime-rich water. This may form in caves or at springs, for example.

Tuff: a deposit or rock made of compacted volcanic ash.

Tundra: a treeless landscape characterised by the presence of grasses and small shrubs that today is characteristic of the arctic and sub-arctic regions.

Tundra-steppe: an alternative term for *Mammoth Steppe*.

Tunnel valley: a valley formed through erosion by a subglacial river, which in some cases flowed upslope towards the ice front.

Tusks: the enlarged upper *incisor* teeth of mammoths and elephants.

Type fossil: see *holotype*.

Type locality: the place where a geological unit (e.g. a *formation*) was first identified and described, and which is used to define that unit.

Type site: in archaeology, the site where an archaeological *culture*, *industry* or *tradition* was first identified and described, and which gives its name to that culture, industry or tradition. For example the *Creswellian* industry (or culture) is named after Creswell Crags on the Nottinghamshire–Derbyshire border.

Ulna (plural **ulnae**): one of the two bones of the lower arm (with *radius*).

Ultrafiltration: a relatively new procedure for the removal of contaminants from samples of ancient bone to be dated by the *radiocarbon* method. Such contaminants may be of a different radiocarbon age to the bone and if not removed may influence results to give erroneous dates (usually too young). Ultrafiltration is particularly important for samples of an age near the limits of the radiocarbon method, i.e. *Upper Palaeolithic* samples.

Unconformity: in geology, an erosion surface separating a unit of younger strata from a unit of older strata.

Ungulate: a hoofed *mammal*.

Unifacial: with regard to stone tools, unifacial means an artefact worked on only one side. Compare *bifacial*.

Uplift: see *istostatic uplift*.

Upper Palaeolithic: The last of the three major subdivisions of the *Palaeolithic,* dated in Europe (including Britain) to between about 45,000 and 11,400 years ago. Characterised by the extensive use of *blade* technology, the proliferation of regional *traditions* and *cultures*, and developments such as the widespread use of bone and antler for tool making, personal ornament, art, burial, campsites with elaborate structures, etc., all of which are associated with the colonisation of Europe by *Homo sapiens*. In Britain, the period is further subdivided into the *Early Upper Palaeolithic, Late Upper Palaeolithic, Final Upper Palaeolithic* and *Terminal Upper Palaeolithic.*

Upper Pleistocene: the last of the three subdivisions of the *Pleistocene* geological *epoch*, dating from approximately 127,000 BP (the beginning of the *Last Interglacial*) to about 11,400 BP (the beginning of the present interglacial or *Holocene*).

Uranium series dating: a *radiometric* dating technique that is based on the radioactive decay of *isotopes* of uranium. The method is used to date rocks rich in calcium carbonate, such as those deposited by seepage in limestone caves (*speleothems*) or around lime-rich springs. The parent uranium isotopes are soluable in water whereas the daughter products are not. This means that only the parent isotopes are present in the water that seeps into the cave, for example. However, once the calcium carbonate with uranium impurities is precipitated as a speleothem the radioactive clock is set going and the daughter products start to build up. The isotopes are measured by counting their alpha emissions or by using thermal ionising mass spectrometry (TIMS), the latter providing much greater precision. Uranium series dating is most effective in the range of about 500,000–50,000 years ago, so it is particularly useful for dating *Lower* and *Middle Palaeolithic* deposits in limestone caves.

Use-wear analysis: the study of the surfaces of stone tools for traces of alteration resulting from use. As these traces often consist of polish and striations that are only visible under a microscope, this technique is also known as *microwear analysis*. In order to attempt to determine the use of ancient stone tools from use wear, modern replicas are made and used for different purposes, for example cutting meat or carving wood. The patterns of wear produced on the modern artefacts are then compared to the patterns found on the ancient artefacts. Use wear analysis can only be carried out successfully on tools in near 'mint condition', which are a small minority. See also *residue analysis*.

Varves: layers of sediment on the floors of lakes (mainly) that consist of couplets of summer and winter layers which together comprise a varve. The summer and winter layers differ because of seasonal changes in sediment input to the lake or changes in biogenic production. The varves can be counted, somewhat in the manner of annual layers in *ice sheets* (see *ice cores*) or tree rings, and thus dated. Although of limited application, varve chronology applied to proglacial lakes in Scandinavia was used to provide the first reliable date for the end of the *Ice Age*, and more recently varves have been used for the *calibration* of *radiocarbon* determinations beyond the range of *dendrochronology*. 'Varv' is Swedish for layer.

Vegetational climax: see *climax community.*

Ventral: in stone-tool making, the 'bottom' surface of a *flake* (or *blade*). This corresponds to the face of the flake where it was detached from the *core* and will often display features such as a *bulb of percussion* or *ripples*. Compare *dorsal*.

Vertebra (plural **vertebrae**): an individual bone of the spinal column. See also *cervical vertebrae, thoracic vertebrae, lumbar vertebrae* and *caudal vertebrae.*

Vertebrate: an organism with a backbone.

Vole Clock: a *biostratigraphic* method of dating deposits by means of evolutionary changes in the small rodent *fossils* found within them. Such small *mammals* lead short, energetic lives and the evolutionary changes wrought by *natural selection* consequently occur quickly. A particularly important evolutionary transition is that from Mimomys savini, an extinct *species* of water vole, to its lineal descendent Arvicola terrestris cantiana (also now extinct). This transition, which is reflected particularly in the very different teeth of the two species, took place some time between 600,000 and 500,000 years ago, as has been established from deposits in Germany and Holland which can be dated by other methods. In Britain the vole clock has been applied in particular to the important pre-*Anglian* sites at Pakefield, in the 'Mimomys zone', and Boxgrove, in the 'Arvicola zone'.

Weichselian Glacial: the name given in continental northwest Europe to the *Last Glacial*, about 115,000–11,400 BP. Corresponds to the *Devensian* in Britain, the *Würm Glacial* in the Alps and the Wisconsinan in North America.

Wet sieving: a method of extracting organic material such as seeds, *molluscs*, insects and the bones of small *vertebrates* from a sample of deposit. Similar in purpose to *flotation* but different in method. The sample is put into a sieve, lumps of soil are broken up and large stones removed by hand, and then the deposit is washed with water so that the organic material is trapped in the sieve while the soil particles are washed away.

Windermere Interstadial: name given in Britain to the *interstadial* towards the end of the *Devensian* stage, between the *Late Glacial Maximum* and the *Loch Lomond Stadial*. Dated from about 15,400–12,900 BP. Called the *Bølling-Allerød Interstadial* in continental northwest Europe, and the *Lateglacial Interstadial* more generally. In Britain the *Late Upper Palaeolithic* archaeological period is largely contemporary with the earlier part of the Winderemere Interstadial and the *Final Upper Palaeolithic* with the later part.

Wisent: an alternative name for the European *bison*.

Wolstonian stage: originally 'Wolstonian Glacial' was the name given to the British equivalent of the *Riss Glacial* of the Alps when the simple four *glacial* scheme developed in the Alpine region (Günz–Mindel–Riss–Würm) was used as a model for the whole of Europe (and beyond). It is now realised that the Wolstonian covers the whole period from *MIS 10 to* MIS 6 inclusive (about 364,000–127,000 BP), embracing three cold stages (MIS 10, 8 and 6) and two intervening *interglacials* (MIS 9: 'Purfleet', and MIS 7: 'Aveley'). The idea of a single 'Wolstonian Glacial' is thus misleading but the term 'Wolstonian stage' may still be used to cover the whole of this long period.

Würm Glacial: name given in the Alpine region to the *Last Glacial*, about 115,000–11,400 BP. Corresponds to the *Devensian* in Britain, the *Weichselian* in continental northwestern Europe and the Wisconsinan in North America.

Younger Dryas Stadial: the name given in continental northwest Europe, but now used much more widely, to the last *stadial* event of the *Pleistocene*, from about 12,900–11,400 BP. Follows the *Allerød Interstadiial* and immediately precedes the commencement of the *Holocene* epoch. Also known as the *Lateglacial Stadial* and equivalent to the *Loch Lomond Stadial* in Britain. The evidence for severe climate cooling during this stadial, especially in western Europe, is impressive. There was renewed glacier activity in mountain regions, including western Scotland (the 'Loch Lomond Readvance') and a return of periglacial conditions to many lowland areas. *Coleopteran* evidence suggests that in Britain summer temperatures fell below 10°C, with winter temperatures in the range −15 to −20°C. It is possible that there was a break in the human occupation of Britain during the most severe part of this episode, although some *Terminal Upper Palaeolithic industries* are associated with it. Named after the alpine/tundra flower <u>Dryas</u> octopetala (mountain avens), the pollen of which is characteristic of deposits of this episode. Compare *Older Dryas Stadial*. See also *Heinrich events* and *thermohaline circulation* for mechanisms that may have been involved in triggering this stadial.

Further Reading

General Books About the Ice Age

Andersen, Bjørn G. and Borns, Harold W. Jr. (1997) *The Ice Age World.* Scandinavian University Press, Oslo.

A lavishly illustrated overview, with an emphasis on northern Europe and North America, in three main parts. The first part provides a history of the concept and study of the Ice Age. The second part is a geological history of the Ice Age. The third part is about processes and scientific methods. Now slightly out of date.

Imbrie, John and Imbrie, Katherine Palmer (1979) *Ice Ages: solving the mystery.* Harvard University Press, Cambridge (Mass.) and London.

Now something of a classic, this book provides an exciting history of theories of the Ice Age up to the 1970s, with much fascinating biographical detail about the protagonists. The main theme is the development and ultimate vindication, as a result of the evidence from deep-sea cores, of the 'astronomical theory' of climate change.

Macdougall, Doug (2004) *Frozen Earth: the once and future story of ice ages.* University of California Press, Berkeley, Los Angeles and London.

A lively, non-technical account of the discovery of ice ages and the search for their causes. It covers not just the Quaternary Ice Age but earlier ice ages in the Earth's history, including the Snowball Earth 'episode' early in the history of life.

Wilson, R.C.L., Drury, S.A. and Chapman, J.L. (2000) *The Great Ice Age: climate change and life.* Routledge/The Open University, London.

An undergraduate level textbook in two parts, taking a global perspective. The first part focuses on the evidence for and explanations of climate change. The second part is about ecological change and human origins. Quite technical in places.

Quaternary Environments and Dating

Elias, S.A. (ed) (2006) *Encyclopedia of Quaternary Science (4 vols).* Elsevier Science.

At 2,600 pages and in four volumes, with over 360 articles covering all the major topics in the field, this is the ultimate reference book on Quaternary science, and up-to-date. However, the price is astronomical, so this is strictly a library reference book. Also available on line through ScienceDirect.

Lowe, J.J. and Walker, M.J.C. (1997) *Reconstructing Quaternary Environments* (2nd edition). Pearson Prentice Hall, London.

The standard textbook, comprehensive and authoritative. The book, which is technical but does not assume prior knowledge, covers the full range of evidence and methods used to reconstruct Quaternary landforms and environments, including dating methods. The final chapter puts the methods into practice by providing a detailed account of the last interglacial-glacial cycle (130,000–10,000 BP).

Walker, M. (2005) *Quaternary Dating Methods.* John Wiley, Chichester.

Comprehensive and up-to-date. Technical.

Gaffney, Vincent, Fitch, Simon and Smith, David (2009) *Europe's Lost World: the rediscovery of Doggerland* (CBA Research Report 160). Council for British Archeaology, York.

As we have emphasised repeatedly, Britain was at most times during the Ice Age not an island but a peninsula connected to continental Europe by the vast plains and hills of 'Doggerland'. This lavishly illustrated book tells the extraordinary story of how a team from the University of Birmingham used oil industry technology to reconstruct the surface of Doggerland, now buried beneath tens of meters of sea water and marine sediment. The world of hunter gatherers who inhabited Doggerland is reconstructed. As on land, the extraction of aggregate from the sea bed opens a window on the Ice Age past if proper recording procedures are put in place.

ICE CORES AND ABRUPT CLIMATE CHANGE

Alley, Richard B. (2000) *The Two-Mile Time Machine: ice cores, abrupt climate change and our future.* Princeton University Press, Princeton and Oxford.

Mayewski, Paul Andrew and White, Frank (2002) *The Ice Chronicles: the quest to understand global climate change.* University Press of New England, Hanover and London.

These two books, written for a general audience by scientists at the centre of the work, provide accounts of the cores drilled into the Greenland ice sheet, the startling results obtained and their implications.

Cox, John D. (2005) *Climate Crash: abrupt climate change and what it means for our future.* Joseph Henry Press, Washington D.C.

A racy but well-researched book by a science journalist reviewing the evidence for abrupt climate change during the Quaternary and assessing its significance for the future.

GLACIERS AND GLACIAL ENVIRONMENTS

Anderson, David (2004) *Glacial and Periglacial Environments.* Hodder & Stoughton.

Part of the 'Access to Geography' series. A succinct, clear introduction to glacial and periglacial processes and landforms.

Hambrey, Michael and Alean, Jürg (2004) *Glaciers* (2nd edition). Cambridge University Press, Cambridge.

A beautifully illustrated (but the subject matter helps) survey of everything you ever wanted to know about glaciers and glacial landscapes for the non-specialist reader. Mainly about glaciers today but with some coverage of glacial history.

ICE AGE AND PALAEOLITHIC BRITAIN

Barton, Nick (2005) *Ice Age Britain*. Batsford/English Heritage, London.

A clear, straightforward survey of the human occupation of Britain during the Ice Age, written for the general reader from the perspective of an archaeologist. Part of the excellent English Heritage series of books on the archaeology of Britain.

Jones, R.L. and Keen, D.H. (1993) *Pleistocene Environments in the British Isles.* Chapman and Hall, London.

A scholarly survey of the environmental evidence at an undergraduate textbook level. Now somewhat out of date on some issues.

McNabb, John (2007) *The British Lower Palaeolithic: stones in contention*. Routledge, Abingdon and New York.

The focus of this book is the Clactonian, a Lower Palaeolithic non-handaxe stone tool tradition. However, the book also provides an excellent introduction to the geological, climatological, evolutionary and chronological frameworks for the study of the British Palaeolithic, and an up-to-date survey of hominin occupation from the earliest record up to the apparent abandonment around the end of MIS 7 (c.186 kya). The book exemplifies the use of archaeological and other evidence for reconstructing hominin societies.

Pitts, Michael and Roberts, Mark (1997) *Fairweather Eden: life in Britain half a million years ago as revealed by the excavations at Boxgrove.* Century, London.

A popular account of the story and results of the Boxgrove excavations, putting them in their wider context. It is very difficult to overestimate the importance of Boxgrove for our understanding of the behaviour, capabilities and mind of pre-modern hominins. Here the evidence is presented and discussed in an accessible and lively manner.

Stringer, Chris (2006) *Homo Britannicus.* Allen Lane/Penguin, London.

This very well written and beautifully illustrated book is written by one of the world's leading authorities on human evolution, who is also the director of the Ancient Human Occupation of Britain (AHOB) project. It provides an account for the general reader of the human occupation of Britain during the Ice Age drawing on the exciting results of the AHOB project, including the revolutionary evidence from Pakefield and Happisburgh for the earliest human occupation of Britain (so far). The final chapter explores the broader implications of climate change. There is also a Penguin paperback edition of the book, in a smaller format.

Wymer, John (1999) *The Lower Palaeolithic Occupation of Britain* (2 vols). Wessex Archaeology and English Heritage, Salisbury.

This two volume book (text and maps) by the late John Wymer is a synthesis of the results of the 'English Rivers Palaeolithic Survey'. This was a seven-year project, stimulated by the massive increase in sand and gravel quarrying, which aimed to identify and map "the find-spots of Lower and Middle Palaeolithic artefacts and the deposits containing them in order to demonstrate fully the distribution of known Palaeolithic sites in Britain". It is an essential resource for anybody researching Pleistocene remains in gravel quarries. As well as presenting the evidence from the river valleys, raised beaches and lakeside environments, primarily investigated as a consequence of quarrying, other environments (e.g. caves and uplands) are included to provide comprehensive coverage. Two substantial introductory chapters provide the necessary background and interpretative framework.

Human Evolution

A huge number of books, aimed at both an academic and general readership, have been published on human evolution, with many new titles appearing every year. The following examples, which take different approaches, are amongst those which the student and general reader should find most useful and interesting.

Stringer, Chris and Andrews, Peter (2005) *The Complete World of Human Evolution*. Thames & Hudson/Natural History Museum, London.

As the name suggests, a comprehensive survey. It is also very well illustrated, with numerous fine reconstructions, and up-to-date, including for example a chapter on the recently discovered *Homo floresiensis* from the island of Flores, Indonesia.

Lewin, Roger and Foley, Robert A. (2004) *Principles of Human Evolution* (2nd edition). Blackwell Publishing, Oxford.

This text focuses less on the story of human evolution and more on how it is studied and how the evidence is interpreted. It is very strong on placing human evolution within the context of evolutionary theory (something missing from many accounts).

Lewin, Roger (2004) *Human Evolution: an illustrated introduction* (5th edition). Blackwell Publishing, Oxford.

Covers much the same ground as the previous book, but in a shorter, introductory format.

Lockwood, Charles (2007) *The Human Story: where we came from and how we evolved*. Natural History Museum, London.

Up to date, authoritative, with very clear text and excellent illustrations. Probably the best short introduction.

Dunbar, Robin (2004) *The Human Story: a new history of mankind's evolution*. Faber & Faber, London.

This book offers a radically different perspective on human evolution from the usual 'bones and stones' approach. Dunbar is an evolutionary psychologist and he draws on the concepts and findings of this new discipline to provide a speculative but very stimulating fresh account of human evolution, focussing on the evolution of the mind.

Neanderthals

Stringer, Christopher and Gamble, Clive (1993) *In Search of the Neanderthals: solving the puzzle of human origins*. Thames & Hudson, London.

Still the best and most balanced general account of the Neanderthals, strong on the palaeoanthropological, archaeological and environmental dimensions, if now becoming a little out of date in some respects. Subsequent research on ancient DNA, however, lends additional strength to the central thesis.

Palmer, Douglas (2000) *Neanderthal*. Macmillan/Channel 4 Books, London.

Accompanied a Channel 4 television series and includes many reconstructions which bring the Neanderthals to life. Well-researched, wide-ranging and readable.

van Andel, Tjeerd H. and Davies, William (2003) *Neanderthals and modern humans in the European landscape during the last glaciation: archaeological results of the Stage 3 Project*. McDonald Institute for Archaeological Research, Cambridge.

A collection of academic papers arising from the 'Stage 3 Project', which carried out a detailed study of the climate and environment of Europe during MIS 3 (60,000–25,000 BP). The work exemplifies recent approaches to climate and environmental reconstruction in the light of evidence for rapid climate change, and human responses to such change.

Ice Age Art

The figurative art of the Upper Palaeolithic, which mainly depicts animals, provides a wonderful window on aspects of Ice Age life towards the end of the Pleistocene (from about 35,000 BP), as well as having great appeal to a modern audience. Numerous lavishly illustrated books have been published, on the art as a whole or on individual regions or caves (mainly in France and northern Spain). Until 2003, however, no parietal art (art painted or engraved on cave walls or other surfaces) had been discovered in Britain, although a handful of portable art objects were known. This changed on 14 April of that year with the discovery of cave art at Creswell Crags. Only books published after 2003, therefore, will include the Creswell engravings.

Bahn, Paul G. and Vertut, Jean (1997) *Journey through the Ice Age.* Weidenfield & Nicholson, London.

The best general account in English, with beautiful photographs by Jean Vertut. Due to the date of publication it does not of course include Creswell Crags, although the author, Paul Bahn, was a member of the team which subsequently discovered the art there.

Bahn, Paul (2007) *Cave Art: a guide to the decorated Ice Age caves of Europe.* Francis Lincoln, London.

The first comprehensive guidebook to all the decorated Ice Age caves open to the public, it includes Creswell Crags. Useful short introduction.

Bahn, Paul and Pettitt, Paul (2009) *Britain's Oldest Art: the Ice Age cave art of Creswell Crags.* English Heritage, Swindon

The definitive account of Creswell Crags art, including the archaeological context.

Stone Tool Technology

Lord, John W. (1993) *The Nature and Subsequent Uses of Flint, Volume 1: the basics of lithic technology.* Self published, Brandon.

John Lord is perhaps the most accomplished flint knapper working in Britain. This short 'how to' book is the fruit of many years of practical experience and is all one needs (minus the raw materials) to get started. It is no substitute, however, for watching John Lord in action. To do so gives one a very forcible impression of the skills of even the most so-called primitive of our ancestors.

Schick, Kathy D. and Toth, Nicholas (1993) *Making Silent Stones Speak: human evolution and the dawn of technology.* Weidenfeld & Nicholson, London. (Paperback edition 1995, Phoenix)

A wide-ranging survey of Palaeolithic stone tools with a global perspective. Covers how the tools were made, who made them, what they were used for and how they are interpreted. Includes evidence from replications, experiments, ethnographic parallels and microwear analysis. It lives up to its title and provides a surprisingly good read for those who mistakenly believe that the study of stone tools must be terminally dull.

Ice Age Mammals

The plants and invertebrate animals of the Ice Age evolved very little. They are still around today, albeit, due to climate change, in different places. Many mammals, however, including humans, underwent great evolutionary change (and its complement, extinction) during the Ice Age. While one is almost swamped by the plethora of books about human evolution, accessible books about the evolution and fate of other Ice Age mammals are a rarity.

Lister, Adrian and Bahn, Paul (2007) *Mammoths: giants of the Ice Age.* Francis Lincoln, London.

Covers all aspects of mammoths – discovery, evolution, behaviour, role in human culture, extinction – with numerous colour illustrations and reconstructions. This is a new and substantially updated edition of an earlier book published in 1994. All the major new discoveries and interpretations are incorporated, including the discovery in May 2007, in northern Siberia, of 'Lyuba', a frozen mammoth calf and "arguably the most complete carcass of a woolly mammoth, or indeed of any prehistoric animal, ever discovered". Gripping.

Sutcliffe, A.J. (1985) *On the Track of Ice Age Mammals*. British Museum (Natural History), London.

Obviously now rather out of date.

Kurtén, Björn (2007) *Pleistocene Mammals of Europe*. Aldine Transaction, New Brunswick and London.

A paperback reprint of a book first published in 1968, so obviously very out of date in some respects. However, it provides a comprehensive survey of all the Pleistocene species of Europe.

Agustí, Jordi and Antón, Mauricio (2002) *Mammoths, Sabertooths, and Hominids: 65 million years of mammalian evolution in Europe*. Columbia University Press, New York.

While covering the whole period (Cenozoic) since the extinction of the dinosaurs, the final chapter is about the Pleistocene. Many fine drawings of mammal skeletons and reconstructions by Antón.

Prothero, Donald R. (2006) *After the Dinosaurs: the age of mammals*. Indiana University Press, Bloomington and Indianapolis.

The same chronological span as Augustí and Antón but with global coverage.

Schreve, D.C. (ed.) (2004) *The Quaternary Mammals of Southern and Eastern England Field Guide*. Quaternary Research Association, London.

One of the Quaternary Research Association's 'Field Guide' series. Much more technical and detailed than the above books but more up to date and specific to England. It mainly comprises descriptions and interpretations of the mammalian fauna from key sites and localities, set in their stratigraphic context, and is not a general survey as such. Available through the Quaternary Research Association (see 'useful organisations').

Guthrie, R. Dale (1990) *Frozen Fauna of the Mammoth Steppe: the story of Blue Babe*

This book, which is something of a minor classic, is mainly a detailed account of the discovery, excavation and analysis of a 36,000-year-old bison mummy found in an Alaskan gold mine and nicknamed 'Blue Babe'. However it also tells the fascinating wider story of frozen, mummified Ice Age mammals and develops and discusses the concept of the 'Mammoth Steppe'.

Guidance and Policy Documents Relating to the Conservation of Geological and Archaeological Remains

English Nature, Quarry Products Association and Silica and Moulding Sands Association (2003) *Geodiversity and the minerals industry – conserving our geological heritage*.

This booklet defines geodiversity, explains its importance, and provides practical advice on achieving conservation and good practice in a quarrying context.

Office of the Deputy Prime Minister (2005) *Planning Policy Statement 9: biodiversity and geological conservation*.

The Government's national policy. It stresses the need for sustainable development and for conservation of wildlife and geology.

English Heritage (1998) *Identifying and protecting Palaeolithic remains – archaeological guidance for planning authorities and developers*.

With an emphasis on sand and gravel extraction sites, this leaflet explains the nature and importance of Palaeolithic sites and how they are protected through the planning process (PPG 16) following the same principles as sites of later periods. Obtainable as free PDF download from the English Heritage website, www.english-heritage.org.uk

Pettitt, Paul, Gamble, Clive and Last, Jonathan (eds) (2008) *Research and conservation framework for the British Palaeolithic.* English Heritage and The Prehistoric Society.

Based on consultation with a broad range of archaeologists and Quaternary scientists, this booklet sets out research and conservation goals for the British Palaeolithic for the five or so years following its publication. As well as defining the academic research themes that need addressing, such as colonisation and settlement processes, the booklet also deals with such issues as education and 'outreach', professional training, curation and conservation, and the enhancement of collections and records.

Department of the Environment (1990) *Planning Policy Guidance Note 16: archaeology and planning.*

The Government's national policy on how archaeological remains should be protected through the planning process.

Institute of Field Archaeologists (2001) *Standard and guidance for archaeological desk-based assessment* (revised edition)

Institute of Field Archaeologists (2001) *Standard and guidance for archaeological watching brief* (revised edition)

Although these two documents were not written with Pleistocene/Palaeolithic deposits in mind, suitably adapted to purpose they provide much useful guidance. Both can be downloaded as PDF files free of charge from the IFA's website, www.archaeologist.net.

Jones A.P., Tucker M.E. and Hart J. (eds) (1999) *The Description and Analysis of Quaternary Stratigraphic Field Sections.* Quaternary Research Association Technical Guide No. 7 (reprinted 2007). Quaternary Research Association, London.

The guide is in two parts. The first part provides background and detailed guidelines on the recording of sections. The second part presents case studies from around the world. Available through the Quaternary Research Association (see 'useful organisations').

JOURNALS AND PUBLICATION SERIES

There are so many useful journal papers and articles relevant to the British Palaeolithic and Pleistocene that no attempt is made to list these here. Instead, a list is provided below of *some* of the principal journals that regularly carry relevant articles. In many cases the contents of journal issues are available on line together with abstracts of articles and papers. In some cases full articles may be downloaded.

Not included is the wide range of general science publications, such as the prestigious British journal *Nature* and its US counterpart *Science,* or the more popular *New Scientist* and *Scientific American*, which will carry papers and articles about all aspects of Quaternary science if they are of sufficient international importance.

QUATERNARY SCIENCE/GEOLOGY

Journal of Quaternary Science (JQS)

(the journal of the Quaternary Research Association, see 'useful organisations' below) www.interscience.wiley.com/journal/jqs

Field Guide and Technical Guide series of the QRA

(very useful for the fieldworker, see 'useful organisations' below)

Quaternary Science Reviews (QSR)

(an international multidisciplinary research and review journal)

www.elsevier.com/locate/qsr

Proceedings of the Geologists' Association (PGA)

(the journal of the Geologists' Association, see 'useful organisations' below)

www.sciencedirect.com/science/journal/00167878

ARCHAEOLOGY

Proceedings of the Prehistoric Society (PPS)

(the journal of the Prehistoric Society, see 'useful organisations' below)

www.ucl.ac.uk/prehistoric/index.html

Antiquity

(a quarterly, peer-reviewed journal of world archaeology. The useful website provides titles and abstracts of contents, including all back numbers, 'open access' pages and details of events. The journal is available on line to subscribers.)

www.antiquity.ac.uk

Lithics

(the journal of the Lithic Studies Society, see 'useful organisations' below)

www.britarch.ac.uk/lithics

British Archaeology

(The popular bimonthly magazine of the Council for British Archaeology, see 'useful organisations' below)

www.britarch.ac.uk/ba

Current Archaeology

(A popular archaeology magazine with a British focus)

www.archaeology.co.uk

Useful Organisations and Websites

The list below is by no means comprehensive. In particular there are too many universities with departments dealing with Quaternary science and Palaeolithic archaeology, and museums with important archaeological and fossil collections, to include these here. However, a local museum often provides a good starting point for getting advice on finds.

Whilst every effort has been made to ensure that contact details and website addresses are correct at the time of writing (2008), these may of course change.

International Union for Quaternary Research (INQUA)

Web: www.inqua.tcd.ie
INQUA was founded in 1928 by a group of scientists seeking to improve understanding of environmental change during the glacial ages through interdisciplinary research. Today it has more than 35 member countries and is a full Scientific Union member of the International Council for Science. The work of INQUA is largely carried out through the activities of its commissions and committees, and it holds quadrennial international congresses, which bring together as many as 1000 scientists from around the world.

Quaternary Research Association (QRA)

Web: qra.org.uk
The QRA is an organisation comprising archaeologists, botanists, civil engineers, geographers, geologists, soil scientists, zoologists and others interested in research into the problems of the Quaternary. The majority of the membership is British. It organises meetings, conferences and field trips. Members receive regular circulars and the *Quaternary Newsletter*. The journal of the QRA is the *Journal of Quaternary Science* (JQS). Other publications of the QRA include its 'Field Guides' and 'Technical Guides' series (very useful for 'Pleistocene prospection'). Refer to the website for current information on officers of the QRA, membership and how to obtain publications.

Quarry Products Association (QPA)

Gillingham House
38-44 Gillingham Street
London
SW1V 1HU
Tel: 020 7963 8000
Email: info@qpa.org
Web: ww.qpa.org
The Quarry Products Association is the trade association for the minerals industry and the vast majority of quarry companies involved in aggregates extraction in the UK are members. The QPA represents its members interests on such issues as policy and planning, which includes archaeological and geological conservation. Statements on the QPA's position on heritage (archaeology) and geology are mainly to be found in the 'Sustainability' pages of its website. The main QPA website also provides links to its other websites, which include 'Safe Quarry' and, for children, 'The Virtual Quarry'.

English Heritage

Customer Services Department
PO Box 569
Swindon

SN2 2YP
Tel 0870 333 1181
Email: customers@english-heritage.org.uk
Web: www.english-heritage.org.uk
English Heritage (officially the Historic Buildings and Monuments Commission for England) is the Government's statutory advisor on the historic environment. It main role is to conserve the historic environment, broaden public access to heritage and increase understanding of the past. At the time of writing English Heritage is one of a number of bodies that distribute grants from the Aggregates Levy Sustainability Fund (ALSF). This fund was set up to tackle problems in areas affected by the extraction of aggregates. The fund has helped numerous projects involved with Palaeolithic remains found in quarries. To find out more about these projects follow the ALSF links on the website.

Natural England

Northminster House
Peterborough
PE1 1UA
Tel: 0845 600 3078
Email: enquiries@naturalengland.org.uk
Web: www.naturalengland.org.uk
Natural England is the Government's statutory advisor on the natural environment, which includes geology. Its main role is to conserve and enhance the natural environment, and to promote enjoyment and understanding of the natural environment and its sustainable use and management. It was formed in 2006 from a merger of English Nature with the Countryside Agency. At the time of writing Natural England is one of a number of bodies that distribute grants from the Aggregates Levy Sustainability Fund (ALSF). This fund was set up to tackle problems in areas affected by the extraction of aggregates.

British Geological Survey (BGS)

Kingsley Dunham Centre
Keyworth
Nottingham
NG12 5GG
Tel: 0115 936 3143
Email: Enquiries@bgs.ac.uk
Web: www.bgs.ac.uk
The BGS is the UK's national geological survey organisation, responsible for advising the Government on all aspects of geoscience, and a centre for Earth Science information and expertise. It is a component organisation of the Natural Environment Research Council (NERC). BGS maps provide national coverage of both 'superficial deposits' ('drift' – i.e. Quaternary deposits) and 'bedrock' ('solid') geology at a variety of scales in both hardcopy and digital formats. The 1:50000 scale maps are accompanied by *Geological Memoirs* which contain detailed information on the structure, stratigraphy and palaeontology of the district, together with extensive references and bibliographies. These are an essential resource for 'Pleistocene prospection'. The BGS also produces a wide range of other relevant reports and publications, including historical maps, and has a database of boreholes. Full details on the BGS website.

Geologists' Association (GA)

Burlington House
Piccadilly
London
W1J 0DU
Tel: 020 7434 9298
Email: geol.assoc@btinternet.com

Web: www.geologists.org.uk
The Geologists' Association exists for all geologists and earth scientists, whether professional or amateur. The GA runs field trips, lectures and events that are open to all, and it publishes a news magazine, a 'what's on' leaflet and a professional scientific journal (its 'Proceedings'). It also has a junior division, 'Rockwatch', for young geologists. Through the 'Local Groups' page on the GA website, links are available to affiliated local groups and other relevant organisations.

AHOB (Ancient Human Occupation of Britain)

Web: www.nhm.ac.uk/hosted_sites/ahob
The AHOB project is a major research project, funded by the Leverhulme Trust, which is investigating the pattern of the human occupation of Britain during the Ice Age. The first stage of the project (AHOB 1) ran from 2001 to 2006, and the results are summarised for the general reader in Chris Stringer's book *Homo Britannicus* (see further reading). The second stage of the project (AHOB 2) is due to run until 2010 and has the subtitle 'Ancient Britain in its European Context'. The AHOB website includes various resources including a picture library and a time chart.

Creswell Crags ("Home of the Ice Age hunter")

Creswell Heritage Trust
Creswell Crags Visitor Centre
Crags Road
Welbeck
Worksop
Notts
S80 3LH
Tel: 01909 720378
Email: info@creswell-crags.org.uk
Web: www.creswell-crags.org.uk
Creswell Crags is a limestone gorge on the Derbyshire/ Nottinghamshire border, honeycombed with caves and smaller fissures. Stone tools and animal bones document Middle and Upper Palaeolithic occupation, while in 2003 engraved art was discovered for the first time in one of the caves, Church Hole. There is a visitor centre with a small Ice Age exhibition – at the time of writing plans for a much larger visitor centre are underway. The website provides details of the gorge, visitor and educational facilities, and events. It also provides a link to *Virtually the Ice Age*, the centre's excellent interactive website.

The Prehistoric Society

Institute of Archaeology
University College London
31-34 Gordon Square
London
WC1H 0PY
Email: Prehistoric@ucl.ac.uk
Web: www.ucl.ac.uk/prehistoric
The Prehistoric Society's interests are world wide and extend from the earliest human origins to the emergence of written records. Many Palaeolithic archaeologists are members and the Society's journal, *Proceedings of the Prehistoric Society* (PPS) carries important reports and papers on Palaeolithic archaeology in the UK and abroad. The Society organises lectures, study tours and events, and the newsletter *PAST* is available to members. Membership is open to all.

The Lithic Studies Society

Email: info@lithics.org
Web: www.britarch.ac.uk/lithics

The Society is concerned with the study of lithics (stone tools) of all periods from all parts of the world. However, Palaeolithic artefacts from the UK are a major focus of interest. The Society holds lectures and day meetings, and organises conferences and field trips. It publishes an annual journal, *Lithics*, as well as occasional papers. The Society promotes the highest standards of lithics research and reporting. Membership is open to all.

Association of Local Government Archaeological Officers (ALGAO)

Tel: 019755 64071
Email: admin@algao.org.uk
Web: www.algao.org.uk
The association provides a forum representing archaeologists working for local authorities and national parks throughout the UK. Its members are senior professional archaeologists employed by local authorities to provide advice on archaeological conservation and management (including with respect to mineral extraction). The website provides a full listing of the relevant local authorities, with names and contact details for the archaeological officers. This source may also be used to find details of the archaeological Sites and Monuments Records (SMRs)/Historic Environment Records (HERs) for each area.

Council for British Archaeology (CBA)

St. Mary's House
66 Bootham
York
YO30 7BZ
Tel: 01904 671417
Email: info@britarch.ac.uk
Web: www.britarch.ac.uk
The CBA is an educational charity working throughout the UK to involve people in archaeology and to promote the appreciation and care of the historic environment. Affiliated to the CBA are eleven regional groups, and its junior division, the Young Archaeologists' Club. The CBA publishes a popular magazine, *British Archaeology* as well as a range of scholarly publications. The website provides access and links to numerous online resources.

Institute *for* Archaeologists (IfA)

SHES
Whiteknights
University of Reading
PO Box 227
Reading
RG6 6AB
Tel: 0118 378 6446
Email: admin@archaeologists.net
Web: www.archaeologists.net
The *Institute for Archaeologists* (IFA) is the professional organisation for practicing archaeologists. In 2008 it changed its name from the *Institute of Field Archaeologists* to reflect its more encompassing character. It promotes standards and ethics in archaeological practice. Many of its useful publications on professional and technical matters will be found under the old name.

Appendix 1: Example of Sediment Description Sheet

Site Name:		Site Code:
Grid Ref:		County:
Project No:		Date:
Area No. or description:		
Sheet No:		Drawing No(s):

Height/ Depth	Thickness	Bed number	**Description** (e.g. lithology, grain sizes, sorting, structure, colour, boundaries, palaeocurrent direction)	**Comment** (e.g. interpretation, fossil content)	Photographs	Samples

Note: this sheet is intended for use by non-specialists. Record the information you feel able to provide. Where possible sections should be logged by qualified specialists, who will produce more detailed logs.

APPENDIX 2: EXAMPLE OF SAMPLE RECORD SHEET

Site Name:	
Grid Ref:	County:
Project No:	Date:
Area No. or description:	

*Site Code:	*Sample No:
*Context No:	*Sub-sample No:
Sample volume:	Taken by:

*These must also be written on the sample bag

Type of sample (e.g. bulk/waterlogged):
Purpose of sample:
Related samples (Nos):
Sediment stratigraphy:
Depth of sample (or height AOD):
Photo and/or drawing Nos:
Sediment description:
Sieved? Yes/No Mesh size:
Comment:

Illustration Sources and Credits

Every effort has been made to contact all copyright holders. The authors and publishers would be pleased to correct in future editions any errors brought to their attention.

Fig. 1: A quarry man and specialist inspect fossil remains uncovered by quarrying (NIAN)

Fig. 2: The British landscape during a glacial period (Nick Arber, © Norwich Castle Museum and Art Gallery)

Fig. 3: The British landscape during an interglacial period (Nick Arber, © Norwich Castle Museum and Art Gallery)

Fig. 4: The place of the Quaternary period in a geological time scale (Nigel Dodds and Helen Moulden, © Birmingham Archaeology)

Fig. 5: Three types of variation in the Earth's orbit (Nigel Dodds, © Birmingham Archaeology)

Fig. 6: Britain as part of the European landmass (Nigel Dodds, redrawn from image supplied by Amgueddfa Genedlaethol Cymru National Museum Wales)

Fig. 7: A bed of foraminifera and a foram magnified 2000x (By courtesy of Alix Cage, University of St Andrews)

Fig. 8: Marine oxygen isotope curve (NIAN)

Fig. 9: A profile of temperatures in central Greenland (Nigel Dodds, after National Research Council 2002 *Abrupt Climate Change: Inevitable Surprises*. National Academies Press)

Fig. 10: The great Anglian ice sheet (© AHOB, reproduced with permission from Chris Stringer)

Fig. 11: A table showing the later stages of the Quaternary sequence in Britain (Nigel Dodds, © Birmingham Archaeology)

Fig. 12: Ice Age environments: a braided river flowing through a tundra landscape (Jenni Chambers)

Fig. 13: Ice Age environments: deciduous woodland (By courtesy of Kirsty Nichol, Birmingham Archaeology)

Fig. 14: Ice Age environments: chalk downland (By courtesy of Kirsty Nichol, Birmingham Archaeology)

Fig. 15: Six-stage developmental model of river terrace formation (David Bridgland 2000 "River terrace systems in north-west Europe: an archive of environmental change, uplift and early human occupation" in *Quaternary Science Reviews* 19, 1293 – 1303. Reproduced with kind permission of the author)

Fig. 16: An idealised section through the Middle Trent terrace 'staircase' (Dave Bridgland, Andy Howard, Mark White, Tom White. From Bridgland, D.R., Howard, A.J., White, M.J. & White, T.S. 2006 *The Trent Valley: Archaeology and Landscape of the Ice Age*. University of Durham Press, p. 15. Reproduced with kind permission from Andy Howard)

Fig. 17: Whitemoor Haye Woolly Rhino (© Birmingham Archaeology)

Fig. 18: An artist's reconstruction of 'Lucy', an early hominin from Africa (© Natural History Museum, London)

Fig. 19: A representative sample of early stone tools from Africa (© Natural History Museum, London)

Fig. 20: Australopithecus and various species of the genus *Homo* (© Natural History Museum, London)

ILLUSTRATION SOURCES AND CREDITS

Fig. 21: Reconstruction of a *Homo erectus* group (© Natural History Museum, London)

Fig. 22: Hoxne handaxe (Volume XIII of 'Archaeologia: or, Miscellaneous Tracts relating to Antiquity' published by the Society of Antiquaries of London, 1807)

Fig. 23: Diagram representing human evolution (Nigel Dodds, after Foley, R. and Lahr, M. 2004 "Human Evolution Writ Small" in *Nature*, v. 431, No. 7012, p. 1043)

Fig. 24: A partial child's skull from Gran Dolina, Atapuerca (Photograph by Jose-Manuel Benito Álvarez. This image remains in the Public Domain)

Fig. 25: Some of the humanly-struck flint flakes from Pakefield (© Natural History Museum, London)

Fig. 26: A flint handaxe from Boxgrove (© Boxgrove Project, with thanks to Mark Roberts and Matthew Pope)

Fig. 27: Scatter of flint debris excavated at Boxgrove (© Boxgrove Project, with thanks to Mark Roberts and Matthew Pope)

Fig. 28: An artist's reconstruction of a group of *Homo heidelbergensis* butchering a rhinoceros at Boxgrove (© Natural History Museum, London)

Fig. 29: A wooden spear from Clacton-on-Sea, Essex (© Natural History Museum, London)

Fig. 30: The reconstructed course of the Bytham River (Nigel Dodds, © Birmingham Archaeology)

Fig. 31: A reconstruction of the extinct straight-tusked elephant *Palaeoloxodon antiquus* (Zdenek Burian, © Jiri Hochman)

Fig. 32: A quartzite scraper, handaxe and chopper from Waverley Wood, Warwickshire (Terry Hardaker, from Keen, D.H., Hardaker, T. and Lang, A.T.O. "A Lower Paleolithic industry from the Cromerian (MIS 13) Baginton Formation of Waverley Wood and Wood Farm Pits, Bubbenhall, Warwickshire, UK" in *Journal of Quaternary Science*, vol. 21, Issue 5, pp. 457—470, Figs 7a, 8a and b. With thanks to Terry Hardaker.)

Fig. 33: Handaxes from Waverley Wood (Warwickshire Museum, reproduced by kind permission of Warwickshire Museum Service)

Fig. 34: An artist's reconstruction of a Neanderthal female (© Natural History Museum, London)

Fig. 35: Neanderthal (right) and modern human skulls (left) (© Natural History Museum, London)

Fig. 36: An artist's reconstruction of the 'family' at Atapuerca (Mauricio Anton, © Mauricio Anton)

Fig. 37: The partial human skull from Barnfield Pit, Swanscombe, Kent (© Natural History Museum, London)

Fig. 38: Middle Paleolithic flakes (© Natural History Museum, London)

Fig. 39: A map of Britain/'Doggerland' and the 'Channel River' system (© AHOB, reproduced with permission from Chris Stringer)

Fig. 40: Mammoth steppe (Nigel Dodds, adapted from Guthrie, R.D. 1990 *Frozen Fauna of the Mammoth Steppe*. University of Chicago, Fig. 9.12)

Fig. 41: A flint handaxe from Lynford Quarry, Norfolk (Norfolk Archaeology Service with thanks to Bill Boismier)

Fig.42: An adult mammoth tusk under excavation at Lynford Quarry, Norfolk (Norfolk Archaeology Service with thanks to Bill Boismier)

Fig. 43: Artist's reconstruction of Neanderthals hunting (© English Heritage Photo Library)

Fig. 44: The colonisation of the world by modern humans (*Homo sapiens*) (Nigel Dodds, after Scarre, C. 2005 *The Human Past*. Thames and Hudson, London)

Fig. 45: A statuette sculpted in mammoth ivory (Photographer Thomas Stephan, © Ulmer Museum, Germany, with thanks to Kurt Wehrberger)

Fig. 46: Artist's reconstruction of the burial of the 'Red Lady' (© Amgueddfa Genedlaethol Cymru National Museum Wales)

Fig. 47: Late Upper Palaeolithic artefacts (© Natural History Museum, London)

Fig. 48: Bison engraving, Church Hole cave, Creswell Crags (© Creswell Heritage Trust, with thanks to Paul Bahn for providing the image)

Fig. 49: The enigmatic 'tally stick' from Gough's Cave, Somerset (© Natural History Museum, London)

Fig. 50: An aerial view of the Cheddar Gorge, Somerset (Photographer Adrian Pingstone. This image remains in the Public Domain)

Fig. 51: A human skull from Gough's Cave (© Natural History Museum, London)

Fig. 52: Chart of the late glacial period (Nigel Dodds, © Birmingham Archaeology)

Fig. 53: Quarrying methods of the early 19th century (John Linnell, Kensington Gravel Pits 1811-12, © Tate, London 2008)

Fig. 54: Quarrying methods of the early 21st century (© Birmingham Archaeology)

Fig. 55: Specialists examining an exposed section in a quarry and taking samples (NIAN)

Fig. 56: Different sediment types and their sizes, from sand to boulders (By courtesy of Mark Stephens, University of the South Pacific, Fiji)

Fig. 57: Map showing reconstructed routes of early Pleistocene rivers in relation to geological deposits (NIAN, after Rose, J., Moorlock, B.S.P., Hamblin, R. J. O. 2001 "Pre-Anglian fluvial and coastal deposits in Eastern England: lithostratigraphy and palaeoenvironments" in *Quaternary International* 70, pp. 5 – 22)

Fig. 58: Some of the main rock/mineral types found in British river gravels (By courtesy of Mark Stephens, University of the South Pacific, Fiji)

Fig. 59: Some rocks that indicate the former origin of ice flow (By courtesy of Mark Stephens, University of the South Pacific, Fiji)

Fig. 60: Diagram for assessing the roundness of sediment particles (NIAN)

Fig. 61: Diagram for assessing the degree of sorting of sediments (NIAN, after Jones, A.P., Tucker, M.E. and Hart, J.K. (eds) 1999 *The description and analysis of Quaternary Stratigraphic Field Sections*, Quaternary Research Association Technical Guide No. 7, Fig. 3.3.)

Fig. 62: Typical poorly sorted glacial till (By courtesy of Jim Rose, Royal Holloway, University of London)

Fig. 63: A modern glacier (Jenni Chambers)

Fig. 64: Raised beach sediments at Brighton, south England (By courtesy of Danielle Schreve, Royal Holloway, University of London)

Fig. 65: A 'chatter-marked' flint pebble (By courtesy of Mark Stephens, University of the South Pacific, Fiji)

ILLUSTRATION SOURCES AND CREDITS

Fig. 66: Modern (interglacial) meandering river in England (© Simon Ledingham, showing the River Wampool)

Fig. 67: Past interglacial organic river channel (palaeochannel) revealed in an English quarry (By courtesy of Jim Rose, Royal Holloway, University of London)

Fig. 68: Modern braided river in Iceland (Jenni Chambers)

Fig. 69: Ice Age braided river sediments revealed in an English quarry (By courtesy of Jim Rose, Royal Holloway, University of London)

Fig. 70: Buff coloured primary and secondary loess overlying Late Devensian cryoturbated head, Horton, Gower, South Wales (by courtesy of David Case and Hannah Brown, University of the West of England)

Fig. 71: Ice-wedge cast (By courtesy of Jim Rose, Royal Holloway, University of London)

Fig. 72: A section showing both modern and fossil (buried) soil horizons (By courtesy of Mark Stephens, University of the South Pacific, Fiji)

Fig. 73: The frozen carcass of a baby mammoth, nicknamed 'Dima' (© Natural History Museum, London)

Fig. 74: The Hall of the Bulls, Lascaux Cave, France (Image in the Public Domain)

Fig. 75: The main Ice Age mammal localities in Britain (© NIAN, after Antony Sutcliffe 1985 *On the Track of Ice Age Mammals* published by the Natural History Museum)

Fig. 76: The skeleton of a rhinoceros (Helen Moulden, © Birmingham Archaeology, after Cornwall 1956 *Bones for the Archaeologist*)

Fig. 77: Excavating the remains of a fossil aurochs (wild cattle) (By courtesy of Danielle Schreve, Royal Holloway, University of London)

Fig. 78: Vole jaw (By courtesy of Danielle Schreve, Royal Holloway, University of London)

Fig. 79: Side view and biting-surface view of vole molar (By courtesy of Danielle Schreve. With thanks to Phil Crabb at the NHM Photographic Unit for photograph)

Fig. 80: Shrew jaw (By courtesy of Danielle Schreve. With thanks to Phil Crabb at the NHM Photographic Unit for photograph)

Fig. 81: The lower jaw (mandible) of a wolf (By courtesy of Danielle Schreve. With thanks to Phil Crabb at the NHM Photographic Unit for photograph)

Fig. 82: The skull of a hyaena (By courtesy of Danielle Schreve, Royal Holloway, University of London)

Fig.83: The skull of a lion (By courtesy of Danielle Schreve. With thanks to Phil Crabb at the NHM Photographic Unit for photograph)

Fig. 84: The upper jaw of a bear (By courtesy of Danielle Schreve. With thanks to Phil Crabb at the NHM Photographic Unit for photograph)

Fig. 85: Red deer antler (By courtesy of Danielle Schreve. With thanks to Phil Crabb at the NHM Photographic Unit for photograph)

Fig. 86: Bison horns (By courtesy of Danielle Schreve. With thanks to Phil Crabb at the NHM Photographic Unit for photograph)

Fig. 87: The skull of an aurochs (wild cattle) (By courtesy of Danielle Schreve. With thanks to Phil Crabb at the NHM Photographic Unit for photograph)

Fig. 88: Straight-tusked elephant lower tooth (By courtesy of Danielle Schreve. With thanks to Phil Crabb at the NHM Photographic Unit for photograph)

Fig. 89: Mammoth upper tooth (By courtesy of Danielle Schreve. With thanks to Phil Crabb at the NHM Photographic Unit for photograph)

Fig. 90: The increasing number of enamel lamellae (plates) forming the teeth of different mammoth species through time (NIAN, after Lister, A. 1993 "'Gradualistic' evolution: Its interpretation in Quaternary large mammal species" in *Quaternary International*, 19, pp. 77 – 84)

Fig. 91: Woolly rhinoceros upper teeth (By courtesy of Danielle Schreve. With thanks to Phil Crabb at the NHM Photographic Unit for photograph)

Fig. 92: Horse upper teeth (By courtesy of Danielle Schreve. With thanks to Phil Crabb at the NHM Photographic Unit for photograph)

Fig.93: Cut-marked brown bear paw bones (arrows indicate cuts) (By courtesy of Danielle Schreve. With thanks to Phil Crabb at the NHM Photographic Unit for photograph)

Fig. 94: Reindeer bone broken for marrow (arrow indicates impact) (By courtesy of Danielle Schreve, Royal Holloway, University of London)

Fig. 95: Carved mammoth figure (By courtesy of Danielle Schreve, © NHM Photographic Unit)

Fig. 96: The human skeleton showing the principal bones and bone groups (Nigel Dodds, © Birmingham Archaeology)

Fig. 97: Modern distribution of two snail species, *Columella collumella* and *Clausilia pumila*, which lived in Ice Age Britain (NIAN, David Keen)

Fig. 98: Some examples of common gastropod (snail) species (NIAN, some images adapted from Kerney, M.P. and Cameron, R.A.D. 1979 *A Field Guide to the land snails of Britain and North-West Europe*, Collins)

Fig. 99: Some examples of bivalve (mussel and clam) species (NIAN, David Keen)

Fig. 100: Top: Shells exposed in the sands of the Thames. Bottom: The shells after sieving (By courtesy of Danielle Schreve, Royal Holloway, University of London)

Fig. 101: *Corbicula* shells within an aurochs' (wild cattle) skull (Barbara Silva)

Fig. 102: Mollusc evidence for changing environments 200,000 years ago in Buckinghamshire (NIAN, David Keen)

Fig. 103: *Stephanocleonus eruditus* head capsule (By courtesy of Svetlana Kuzmina, Royal Holloway, University of London)

Fig. 104: Diagram of beetle showing principal body parts (Helen Moulden © Birmingham Archaeology)

Fig. 105: Beetle wing cases (By courtesy of Katie Denton & Marc Aymer, Royal Holloway, University of London)

Fig. 106: Chironomid head capsule 1000x (NIAN, by courtesy of Alistar Brown)

Fig. 107: Fossil pine tree stumps buried in peat in Scotland (By courtesy of John Lowe, Royal Holloway, University of London)

Fig. 108: Plant fragments embedded in the teeth of the Whitemoor Haye woolly rhino (NIAN)

Fig. 109: A sedge seed magnified 40x (NIAN, by courtesy of Alistair Brown)

Illustration Sources and Credits

Fig. 110: Pine pollen grain 1000x (Image photographed by Dr N Branch and adapted by Geography Department at Royal Holloway, University of London)

Fig. 111: Dandelion pollen grains 1000x (Image photographed by Dr N Branch and adapted by Geography Department at Royal Holloway, University of London)

Fig. 112: Pollen diagram from Marks Tey (Nigel Dodds, adapted from Turner, C. "The Middle Pleistocene Deposits at Marks Tey, Essex" in *Philosophical Transactions of the Royal Society of London*. Series B, Biological Sciences, Vol. 257, No. 817 (Mar. 5, 1970), pp. 373-437, with thanks to Charles Turner)

Fig. 113: Taking a column sample for pollen analysis using monolith tins (© Birmingham Archaeology)

Fig. 114: A hard hammer (quarzite cobble) being used to detach a large flake from a core (flint nodule) (Jenni Chambers)

Fig. 115: A soft hammer (antler) being used for final flaking of a handaxe (Jenni Chambers)

Fig. 116: Indirect percussion using an antler punch (By courtesy of John Lord)

Fig. 117: Pressure flaking (By courtesy of John Lord)

Fig. 118: A replica Solutrean leaf point (By courtesy of John Lord)

Fig. 119: Map showing the distribution of the major sources of raw materials for stone tools (NIAN)

Fig. 120: A flint flake showing percussion features (NIAN)

Fig. 121: A flint flake stained orangey brown (Jenni Chambers)

Fig. 122: A chert flake, naturally dark brown, showing some orange staining (Jenni Chambers)

Fig. 123: A variety of flakes produced by knapping (NIAN)

Fig. 124: Some common unifacial retouched tool types (NIAN)

Fig. 125: A sequence of flakes being removed from a flint core (Jenni Chambers)

Fig. 126: A flint core showing the percussion features arising from the removal of a sequence of flakes (NIAN)

Fig. 127: A prepared flake (left) and a waste flake (right) (By courtesy of John Lord)

Fig. 128: Refitting (© Boxgrove Project, with thanks to Mark Roberts and Matthew Pope)

Fig. 129: Core reduction method of handaxe production (NIAN, after Gamble 1999.)

Fig. 130: A lightly stained flint handaxe showing some of the principal features (NIAN)

Fig. 131: Examples of handaxes of different materials found in gravel deposits around England (NIAN)

Fig. 132: Handaxe being used for butchery and skinning (Jenni Chambers)

Fig. 133: Some of the common handaxe shapes that have been defined (NIAN, after Wymer, J.J. 1968. *Lower Palaeolithic Archaeology in Britain as represented by the Thames Valley*. John Baker, London)

Fig. 134: A schematic diagram showing the preparation of a Levallois core and detachment of a flake (NIAN, after Mellars, P. 1996. *The Neanderthal legacy: an archaeological perspective from western Europe*. Princeton University Press, Princeton, N.J.)

Fig. 135: A Levallois 'tortoise core' and Levallois flakes (NIAN)

Fig. 136: Hafted point (Nigel Dodds)

Fig. 137: Mousterian tools – point and side-scraper (© Natural History Museum, London)

Fig. 138: Preparation of a blade core (By courtesy of John Lord)

Fig. 139: Blade core showing removal facets (By courtesy of John Lord)

Fig. 140: Blades struck from a single core (By courtesy of John Lord)

Fig. 141: A range of Upper Palaeolithic tools made on blades (© Birmingham Archaeology)

Fig. 142: The uses of Upper Palaeolithic technology: from core, to blade, to blade tool, to antler tool (Helen Moulden, adapted from Scarre, C. 2005 *The Human Past,* London: Thames and Hudson)

Fig. 143: Upper Palaeolithic antler harpoon (© Natural History Museum, London)

Fig. 144: A rib fragment from Robin Hood Cave, Creswell Crags engraved with the image of a horse's head (© British Museum)

Fig. 145: A gamma spectrometer being used in the field to measure the natural radiation present in a sediment (Jenni Chambers)

Fig. 146: Analysing OSL samples in the laboratory (By courtesy of Simon Armitage, Royal Holloway, University of London)

Fig. 147: Looking for handaxes in a quarry (Photographer Elaine A Wakefield, © Wessex Archaeology)

Index

Page numbers in italic refer to illustrations. Entries in the glossary are not included in this index but further information on many topics listed below will be found in the glossary.

Acheulean, 14, 98
Africa, role in human evolution, 23-27; evolution of *Homo sapiens* in, 46
Aggragation (of river valleys), *18*, 19-22
Alaska, 16, 42, 65
Alpine glaciations, 12
Altamira Cave, Spain, 47
Amino acid racemisation (AAR), 29, 110
Amphibians, 69
Ancient Human Occupation of Britain (AHOB) project, 28-29, 166, 174
Andesite, 36, 89, *96*
Anglian Glaciation, 12, *13*, 21, 29, 30, 32, 36
Antarctic ice sheet, 10-11
Art, origins of, 44-46
Arvicola terrestris cantiana (vole), 106
Association of Local Government Archaeological Officers (ALGAO), 175
Astronomical Theory, 7, *7*
Atapuerca, Spain, *27*, 28, 37
Aurignacian, 101
Aurochs, *66*, *68*, 70-72, *72*, *80*
Australia, 66; human colonisation of, 44
Australopithecus, 24, *25*
Aveley Interglacial, 39-40
Avon River (Worcestershire/Warwickshire), 67

Banwell Bone Cave MAZ, 106
Barnfield Pit, Swanscombe, Kent, 37, *39*, 40
Baton, antler, 47, *48*
Bear, *3*, *71*; brown, 17, 42, 69, *75*; cave, 69
Beaver, 69
Beeches Pit, Suffolk, 40
Beeston Sand & Gravel (Middle Trent Valley), *20*, 21
Beetles, 80-83; climate reconstruction using, 82; finding, 82; principal body parts, *81*; wing cases, *82*
Beringia, *42*
Bifaces, *see* handaxes
Biostratigraphy, 16, 73, 105-106

Birds, 47-48, 69
Bison, 17, 28, 30, 42, 47, *49*, 70-72, *72*
Blombos Cave, South Africa, 46
Boar, wild, *3*
Bones, cut-marks, 31, 73, 119; spiral fracture, 73; *see also* vertebrate fossils
Bordes, François, 100
Borehole surveys, 113
Boulder clay, see glacial till
Bovids, 70-72
Boxgrove, West Sussex, 30-32, 30, 31, 32, 34, 36, 59, 76, 97, 106
Braided rivers, 2, 15, 19
Bridgland, David, 19
Brighton, East Sussex, 59
Britain, as a peninsula, 8, importance for human evolution, 23; colonisation by Homo sapiens, 47; human abandonment of, 17, 40, 41
British Geological Survey (BGS), 112, 173
Butchery, 31
Burgos, Spain, 29
Burins, 101
Burials, human, 76-77, 119
Bytham River, 12-13, *13*, 17, 29, 32-36, *33*, 55

Caddis flies, see Tricoptera
Cannibalism, 28, 48, 50
Cat, scimitar-toothed, 28, 69; wild, 69
Caucasus, 26
Cave art, 47, *49*, 66, *66*
Ceprano, Italy, 28
Channel River, 17, 41, *41*
Chatter-marks, 58-59, *59*
Cheddar Gorge, Somerset, 47-50, *49*
Cheddar Points, 47
Chichester, 30
Chile, 66
China, 24
Chironomidae, 82, *82*
Church Hole Cave, Creswell Crags (cave art), 47, *49*

185

Clacton-on-Sea, Essex, 98; wooden spear, 31, 33, 99
Clactonian, 98
Clausilia pumila (mollusc), 78, 78
Climate reconstruction, 9-11; using beetles, 82
Coelodonta antiquitatis, see rhinoceros, woolly
Coleoptera, see beetles
Columella columella (mollusc), 78, 78
Common ancestor (of humans and chimpanzees), 23
Continental drift, 7
Cosmogenic radionuclide dating, 107
Council for British Archaeology (CBA), 175
Crayford, Kent, 40
Creswell Crags, Derbyshire/Nottinghamshire, 47, 50, 101, 174
Creswellian tradition, 47, 48, 50, 101
Cromer Forest-bed Formation, 67
Cromerian, 30
Cromerian Complex, 30n, 59
Coventry, Warwickshire, 34
Crown Inn Beds (Middle Trent Valley), 20, 21

Dating, 3-4; methods, 104-110; numerical dating, 106-108; relative dating, 104-106; see also individual dating methods
Davies, Ray, 20
Deep sea cores, 9-10
Deer, 30, 47, 70-72; fallow, 3, 70; giant, 3, 28, 70; red, 2, 31, 48, 70, 72
Dendrochronology, 108
Denticulate, 92
Desk-based assessments, 112-113
Devensian Glaciation, 12, 41, 47
Diatoms, 87
'Dima' (frozen mammoth carcass), 65
Dmanisi, Republic of Georgia, 26
DNA, ancient, 66; analysis, 73-75
Dog, 48
Doggerland, 41, 42, 42, 47, 50, 63
Dover Strait, flooding of, 8-9,11, 17, 29, 41
Down cutting (of rivers), 19
Dryas octopetala (Mountain Aven), 86
Dunbridge Quarry, Hampshire, 116

Eagle Moor Sand & Gravel (Middle Trent Valley), 20, 21
Early Upper Palaeolithic, 47
Eggington Sand & Gravel (Middle Trent Valley), 20, 21

Elephant, 30, 32, 72; straight-tusked, 2, 3, 16, 28, 34, 35, 72, 73
English Heritage, 111-112, 172
English Rivers Palaeolithic Survey (TERPS), 112-113
Equipment (for watching briefs), 115
Europe, colonisation by modern humans, 44

Field evaluation, 113
Fieldwalking, 113
Final Upper Palaeolithic, 50
Finds, deposition of, 120
Fire, human use of, 39-40
Fish, 69
Flint, 96; formation of, 89
Foraminifera, 9, 9
Fossils, 65-87; exceptional preservation, 65-66; see also vertebrate fossils, mammalian fossils and invertebrate fossils
Fox, 69
Frogs, 69

Gamma spectrometer, 108, 109
Geologists' Association, 120, 173
Geophysical survey, 113
GISP2 ice core, 10-11, 11
Goat's Hole Cave, Paviland, Gower, South Wales, 46, 47
Gough's Cave, Cheddar Gorge, Somerset, 47-50, 48, 49, 50; Mammal Assemblage Zone (MAZ), 106
Glacial cycles, 5-8
Glacial erratics, 58
Glacial till, 57-58, 57
Glaciations, effect of, 8-9, 58
Global colonisation (by modern humans), 44, 44
Goodwood-Slindon Raised Beach, 59
Gran Dolina, Atapuerca, Spain, 27, 28, 29, 106
Gravettian tradition, 101
Greenhouse gases, 7, 11
Greenland ice sheet, 10-11
Greensand chert, 89, 96
Grouse, black, 48
Gulf Stream, 8
Günz Glaciation, 12

Hammers, hard, 88, 88
Hammers, soft, 31, 88, 88
Handaxes, 25-26, 26, 29, 30, 31, 31, 35, 35-36, 36, 95-97; bout coupé, 100; manufacture, 95,

95, 96; materials, 95-96, 96; Neanderthal, 42, 42; shapes and sizes, 97, 97; use 97, 97
Happisburgh Glaciation, 105
Hare, arctic, 48
Harpoons, antler101, 102
Health & Safety, 111
Hedgehog, 69
Heidelberg, Germany, 29
Hemington Gravel (Middle Trent Valley), 20
Hippopotamus, 17, 21, 28, 41
Historic Environment Records (HERs), 112, 119
Hohlenstein-Stadel Cave, Germany (ivory figurine), 45, 46
Holme Pierrepoint Sand & Gravel (Middle Trent Valley), 20
Holocene, 5, 50
Hominid, definition, 23n
Hominins; definition, 23; fossils, 76-77
Homo (genus), 23, 24; possible evolutionary development, 27
Homo antecessor, 27, 28, 29, 29n
Homo erectus, 24-26, 25, 26, 29
Homo ergaster, 24n
Homo georgicus, 26
Homo heidelbergensis, 25, 29, 29n, 30, 31, 34, 35, 44, 76
Homo neanderthalensis, see Neanderthals
Homo sapiens, 23, 24, 25, 38, 39, 41; behaviour, 44; colonisation of Britain, 47; colonisation of world, 44; evolution of, 44-50; interaction with Neanderthals, 46-47; physique, 44
Horse, wild, 17, 30, 31, 42, 47, 48, 72-3, 75
Horton, Gower, South Wales, 62
Hoxne, Suffolk, 12; handaxe from, 26
Hoxnian Interglacial, 12, 31
Human evolution, 23-51
Human skeleton, 77
Hunting, 31; of mammoths, 43, 43; and Gough's Cave, 48
Hyaenas, 28, 30, 31, 32, 70; spotted, 3, 16, 69

Ice Age, importance of, 2-3
Ice cores, 10-11
Ice-wedges, 63
Indonesia, 24
Institute for Archaeologists (IFA), 175
International Union for Quaternary Research (INQUA), 172
Invertebrate fossils, 77-83; importance of, 77; see also beetles and molluscs
Ipswichian Interglacial 12, 17, 21, 41, 105

Italy, 28

Java Man, 24

Keen, David, 1
Kensington Gravel Pits, 52
Knapping, definition, 88-89

La Cotte de St. Brelade, Jersey, 76
Lake District, 36
La Madeleine rock-shelter, France, 47
Language, 26; origins of, 46
Lascaux Cave, France, 47, 66
Last Glacial Maximum (LGM), 47, 50
Late Upper Palaeolithic, 47-50
Leopard, 69
Lemming, 69
Le Moustier, France, 39, 99
Levallois technique, 37-38, 40, 40; 98-99, 98, 99; cores, 98-99, 98; flakes, 98, 99 points, 99, 99
Lion, 2, 16, 28, 30, 31, 69, 71
Lithics Study Society, 120, 174
Loess, 62, 63
Long-blade assemblages, 50
Lower Palaeolithic, 24-36, 95-97
'Lucy' (partial skeleton of Australopithecus afarensis), 24, 24
Lynford Quarry, Norfolk, 42-43, 42, 43
Lynx, 48

Magdalenian tradition, 47, 101
Mammal Assemblage Zones (MAZs), 106
Mammalian fossils 69-77; bovids, 70-72; carnivores, 69; elephants, 72; fossil localities in Britain, 66, 67; hominin, 76-77; perissodactyla (rhinos and horses), 72-3; teeth, 68-75; see also animals by common name
Mammoth, 2, 16, 75, 75; teeth 72, 73, 74; woolly, 42, 43, 43; ivory, 47
Mammoth Steppe, 42, 42
Mammuthus, see Mammoth
Marine Oxygen Isotope Stages (MISs), 9-10, 10, 20-21
Marks Tey, Essex (pollen diagram), 85, 86
Marksworth, Buckinghamshire (mollusc diagram), 80, 80
Mauer, Germany, 29
Medway River, 17
Mental template, 25
Meuse River, 17, 29
Middle Palaeolithic, 39-43, 98-100

Milankovitch, Milutin, 7
Milankovitch Cycles, 7, 11, 20
Mimomys savini (vole), 106
Mindel Glaciation, 12
Mitigation strategies, 112-114
Mole, 69
Molluscs, 78-80; gastropods, 78, 78; bivalves, 76-79, 79; finding and interpreting, 79-80, 79; sampling for, 79-80
Monkey, macaque, 17
Mouse, 69
Mousterian, 39, 99-100, 99; Mousterian of the Acheulean Tradition (MAT), 100; variants, 100
Musk ox, 70, 72
Mustelids, 69
Mutual Climatic Range method, 82

Natural England, 111, 172-173
Neanderthals, 17, 23, 25, 36-44, 37, 38, 76, 100; physique, 36-37; interaction with Homo sapiens, 46-47
Neander Valley, Germany, 36
North Downs, 16
North European Plain, 42
North Sea, 29
Norton Bottoms deposits (Middle Trent Valley), 20, 21
Norton Subcourse, Norfolk, 34
Notch, 92

Olduvai Gorge, Tanzania, 25
Optically Stimulated Luminescence (OSL) dating, 108-109, 109
Otter, 69
Oxygen isotopes, 9
Oxygen Isotope Stages, see Marine Oxygen Isotope Stages

Pakefield, Suffolk, 28-29, 28, 30n, 32, 106; flint flakes from, 28
Pakefield Interglacial, 29, 105
Palaeochannels, 60
Palaeoloxodon antiquus, see elephant, straight-tusked
Paviland Cave, see Goat's Hole Cave
Pebble tools, 2, 25
Peking Man, 24, 26
Penknife points, 50
Piltdown forgery, 76
Planning Policy Guidance Note 16 (PPG 16), 113
Plant remains, 83-87; macrofossils, 83-84, 83, 84; sampling for macrofossils, 84; microfossils, 84-86; see also pollen
Plate tectonics, 7
Pleistocene, definition, 5
'Pleistocene potential', assessment of, 112
Pollen, 84-86, 84; analysis, 86; diagrams, 85, 86; sampling, 86, 87
Pontnewydd Cave, North Wales, 39-40, 76
Portable art, 101
Positive feedback, 8
Prehistoric Society, 120, 174
Prepared core technology (PCT), 37, 98-99
Primary context deposits, 31; recognition of, 118-119
Project advisors, 114
Ptarmigan, 48
Punch, antler, 89, 89

Quarrying, importance for Quaternary science, 1, 13, 15, 17, 52
Quarrying methods and fossil recovery, 52, 52, 53
Quarry Products Association (QPA), 111, 172
Quartzite tools, 35-36, 89, 96
Quaternary, definition, 5, place in geological time scale, 6
Quaternary environments and landscapes, 15-17; glacial environments, 15, 16; interglacial environments, 16-17, 16
Quaternary Research Association (QRA), 113, 120, 172
Quaternary stratigraphy, 13-14, 14

Radiocarbon dating, 106-107; accelerator mass spectrometry (AMS), 107; calibration, 107-108; ultrafiltration, 107
Rafting (of geological deposits), 105
Raised beaches, 58, 59
'Red Lady' of Paviland, see Goat's Hole Cave
Reindeer, 42, 70, 75
Religion, origins of, 44-46
Reporting (on fieldwork), 119-120
Reject piles (in quarrying), 35, 117
Rhine River, 17, 29, 41
Rhinoceros, 17, 28, 30, 31, 32, 67, 72-73; forest, 2; woolly, 2, 16, 20, 21, 40, 42, 74, 75, 83, 84
Riss Glaciation, 12
Rivers; braided, 61; meandering, 60
River terraces, as archives of the Ice Age, 20-21
River terraces, formation, 18, 19-22
River valleys, Pleistocene record from, 17-22

Robin Hood Cave, Creswell Crags (horse engraving) 101, 102
Rodents, 69

St. Acheul, France, 98
Salisbury Plain, 16
Sample record sheets, 118
Sampling, of sediments, 52-3, 53; bulk, 118; for molluscs, 79-80; for pollen, 86, 87
Scavenging, 31
Schöningen, Germany, 31, 40
Scraper, 92
Sea level change, 17
Secondary context sites, 30, 32, 35
Sediments, 52-64; agents of formation, 54; identifying degree of sorting, 57, 57; identifying particle shape, 56, 56; identifying particle size, 54, 55; identifying rock types, 54-56, 55, 56; main types of, 57-64; marine deposits, 58-59, 59; principles of identification and analysis, 54-57; recording of, 63; river deposits, 59-62, 60, 61; sampling, 52-53, 53; wind-blown, 62-63, 62
Shrew, 69, 69
Siberia, 16, 42, 65
Sima del Elefante, Atapuerca, Spain, 28n
Sima de los Huesos, Atapuerca, Spain, 37, 38, 40, 76
Sites and Monuments Records (SMRs), 112, 119
Soils, 63, 64
Solutrean, 101; leaf point, 89, 90
Somme River, 17
South Downs, 30
Spain, 28, 37
Spears, 31, 33, 40, 99
Spear throwers, 101
Stable isotope analysis, 75-76
Stephanocleonus eruditus (weevil), 80, 81
Stone tools, 23-51, 88-103; assemblages, 98; blade technology, 47, 48, 100-101, 100, 101, 102; conchoidal fracture, 89; cores, 91-2, 93; débitage, 92-93, 94; flakes, 90-91, 92; identification, 103; indirect percussion, 89, 89; industries, 98; manufacture, 88-9; microwear, 94; modes of manufacture, 95-101; percussion features, 91; pressure flaking, 89, 89; raw materials 90, 90; refitting 94, 94; retouching, 29, 90-91; staining, 90, 91; see also individual tool types, industries, techniques and materials
Squirrel, 69
Suprainiac fossa, 37
Swan, whooper, 48
Swanscombe, Kent (human fossil), 37, 39, 40, 76

Tally stick, 48, 49
Tame River, 20, 21
Taphonomy, 73
Teeth, animal, 68-74; carnassial, 69
Terminal Upper Palaeolithic, 50
Terrace staircases, 19-22
Terrestrial record, terminology, 11-12
Thames River, 17, 19, 37, 41, 67
Till, see glacial till
Toads, 69
Trent River, 20, 21, 67
Tricoptera (caddis flies), 82-83

Upper Palaeolithic, 44-50, 100-102

Vertebrate fossils, 65-77; identification, 68; methods of excavation, 67; methods of sampling, 67-68; small, 69; see also mammalian fossils
Volcanic aerosols, 11
Vole, 16, 68, 69, 69, 106
Vole Clock, 106

Watching briefs, 113, 114-119
Waverley Wood, Warwickshire, 34-36, 35, 36
Weasel, 69
Weichselian Glaciation, 12
West Runton, Norfolk, 30n
Whitemoor Haye Quarry, Staffordshire (woolly rhinoceros) 20, 21, 53, 83, 84
Wisconsian Glaciation, 12
Wolf, 2, 28, 30, 42, 48, 69, 70
Wymer, John, 112

Younger Dryas, 50, 86

Zoukoudian Cave, China, 26